SUNLIGHT
AND
HEALTH

SUNLIGHT AND HEALTH

Michael J. Lillyquist

DODD, MEAD & COMPANY
New York

Copyright © 1985 by Michael J. Lillyquist
All rights reserved
No part of this book may be reproduced in any form
without permission in writing from the publisher.
Published by Dodd, Mead & Company, Inc.
71 Fifth Avenue, New York, N.Y. 10003
Distributed in Canada by
McClelland and Stewart Limited, Toronto
Manufactured in the United States of America
Designed by Nancy Dale Muldoon
A GAMUT BOOK

Library of Congress Cataloging in Publication Data
Lillyquist, Michael J.
  Sunlight and health.

  Bibliography: p.
  Includes index.
  1. Sun-baths.   2. Solar radiation—Physiological
effect.   3. Health.   I. Title.
RM843.L55   1985   613'.193   84-21084
ISBN 0-396-08482-6

ISBN 0-396-08957-7 {PBK}

For my sons, Michael and Timothy

CONTENTS

ACKNOWLEDGMENTS
ix

INTRODUCTION
1

1 · THERE GOES THE SUN
6

2 · A SUNLIGHT PRIMER
42

3 · THE SUN AS FRIEND
63

4 · THE SUN AS FOE
102

5 · SENSE IN THE SUN
132

6 · THE CIVILIZED SUN
179

APPENDIX
THE SCIENTIFIC STUDY OF LIGHT AND HEALTH
197

NOTES
206

FURTHER READING
211

INDEX
219

ACKNOWLEDGMENTS

I WANT to thank those people who were kind enough to help in the development and completion of this book. I am especially indebted to Cathy Stephenson for encouraging me to tackle this project and for reading, commenting on, and improving the entire manuscript.

Arlene Fox, my mother, drew on her years of experience as an English teacher and refined my words considerably. Tim Lillyquist, my brother, and Liz Gipple read various chapters and offered their helpful comments. Bill MacKinnon was unfailing in his encouragement and, as always, stimulating with his insight.

I would also like to thank the following for their special help: Dr. Daniel Berger of the Temple University School of Medicine; Gerald Cotton with the Air Resources Laboratory of the National Oceanic and Atmospheric Administration; Karin Hazelkorn and Jean Mudron, Coordinator and former Coordinator respectively, of the Sun Awareness Project/University of Arizona Cancer Center; and the librarians of the University of Arizona Libraries. Dr. John Parrish of the Harvard Medical School checked the manuscript for accuracy and gave me his expert advice. Any errors that remain, of course, are my own responsibility.

Finally, my thanks go to Cynthia Vartan and her associates at Dodd, Mead as well as to Michael Larsen and Elizabeth Pomada, all of whom applied their considerable effort and expertise in bringing the project to its present form.

SUNLIGHT
AND
HEALTH

INTRODUCTION

*Give me the splendid silent sun with all his beams
full-dazzling.*
                            —Walt Whitman, *Leaves of Grass*

TAKE a walk in the morning sunshine. Experience every
thought and sensation as you bathe in the light. Yesterday's
dreary mood evaporates as you are buoyed by a wave of opti-
mism. The warmth on your skin is welcome after the chill of
night, and as clothes warm, you move around inside a com-
fortable envelope. Objects are saturated with color and, though
familiar, reveal new aspects. The blue sky and white clouds
form a backdrop for the trees and there is a strong sense of
three-dimensionality. As Francis Bacon said, "The beholding
of the light is itself a more excellent and a fairer thing than all
the uses of it."

As you walk on, a new sensation arises. You feel a mild
tingling on the skin, and you know your pale color will soon
change to a healthy-looking glow.

A cloud drifts across the sun. The warmth fades and as a
breeze begins, you head for home. Minutes ago the world was
bright and clear and warm, but once inside, objects appear
flat and dark, the morning newspaper can't even be read, and
the room is cold. Two well-practiced movements—first to the
light switch, then to the thermostat—and the world again becomes bright and warm.

When we think about sunlight's effects on us, most of us
are aware of only four aspects: light, heat, the darkening of
the skin, and the feeling of well-being it produces. Each of
these can be achieved by other means, however, and the cu-

1

rious person might wonder whether sunlight is necessary at all. We know that it is the source of nearly all the earth's energy and that plants, which constitute the essential base of the food pyramid, cannot survive without it. But could we, personally, do without the sun's rays?

Not really. Sunlight changes us in many ways beyond those already mentioned. A walk in the sun not only encourages us, warms us, tans us, and gives us distinct visual impressions, it also affects numerous other bodily processes. Sunlight causes vitamin D to be synthesized in the skin. The tanning process, while providing cosmetic changes, also furnishes the skin with protection from future overexposure. Sunlight can lower our blood pressure and may even reduce the level of cholesterol in the bloodstream. And as it enters our eyes, sunlight triggers internal processes that affect blood, bones, protein level, and numerous glands and organs. Sunlight can increase our resistance to a number of diseases and even cure some. Without sunlight we lose muscle tone and become weak, sexually apathetic, and depressed. Christoph Hufeland was probably the first to comment on the necessity of sunshine for normal human functioning when, in his 1797 book, *The Art of Prolonging Life*, he observed the devitalized state of persons sequestered over long periods in dungeons. Their poor appearance was due not, he thought, to poor diet and lack of recreation, but to the absence of the sun's beneficence.

Sunshine is essential for optimum health. If it is good, perhaps we should get all we can. But here lies the dilemma. Getting too much causes painful burns, premature wrinkling of the skin, and with increasing frequency in the United States, skin cancer. If this is the case, then perhaps we should stay out of it. The obvious problem is that we can't do both.

If you were to walk in the sunshine with me, the dilemma would become even clearer. I live in Tucson, Arizona, one of the sunniest places in the world. A few steps outside my door and you would realize that this is a land where light is everything. The sky is the deepest blue and the muted gray-greens of the desert are unexpectedly brilliant in the light. Our local tourist bureau, in an effort to turn this abundance into dollars, recently began referring to the city as the "Sunshine Factory" in its promotional literature. The sobriquet is not only inept but unfortunate, since the Tucson sun manufactures another product not so eagerly sought. It produces the highest rate of skin cancer in the nation.

Pity the poor tourists pursuing a holiday in the sun. Paging through the local newspaper, they find daily charts indicating how long a person can stay in the sun before suffering a skin-damaging burn—a burn that, if repeated too frequently, may result in something more serious than pain and peeling. And the published charts reveal a surprisingly brief period of exposure necessary to incur this damage. If the truth be known, a summer's day in Minneapolis delivers three times the sunburning radiation as a sunny winter day in Arizona. What is one to do short of spending winters in Nome and summers in Tierra del Fuego?

As with most paradoxes, our concern over the good/bad nature of sunlight points to a problem with the way we think rather than with sunlight itself. The sun doesn't mind if it's both beneficial and harmful any more than fire is concerned that it warms as well as burns. The trick for us humans is to learn enough about a phenomenon so that we can receive its benefits while avoiding harm.

As we know not to approach fire too closely, we should similarly know enough not to bake endlessly in the sun. But what is sensible sunning? Unfortunately, the average person is not knowledgeable enough about the effects of sunlight on the body, the factors that determine the quality of sunlight, and the individual differences that determine whether a given exposure is harmful or beneficial. And much that we accept as fact about sunlight is indeed myth. The purpose of this book is to provide readers with enough information to make intelligent decisions about their exposure to sunlight.

By exploring the history of fact and opinion concerning sunlight, we can observe the distinct trends that have characterized thought about it. Scientific opinion has recently been swinging toward the negative side of the spectrum. While sunlight was once a panacea used to cure all kinds of illness, it is now perceived in some circles as a poison.

To sun or not to sun, that is the question. An old German proverb cautions that "the funeral coach turns twice as often on the shady side of the street." And Dr. Dubois, a nineteenth-century physician, advised parents to send their ailing children to the country. "Feed them as well as you can, but above all toast them, burn them, roast them in the sun."

Indeed, as Americans took this advice to heart, there was a coincident decline in tuberculosis, cholera, dysentery, typhoid, and numerous other diseases. While modern medicine, improved nutrition, and better sanitation generally get the credit for these victories, most overlook the fact that a new drug was gaining wide acceptance at the same time. That drug was sunshine, and for years Americans have been consuming it at every opportunity, to the point of overdose as it now appears.

The oldest known medical text, the *Nei Ching*, taught that man was a microcosmic representation of the universe, that life and health were a product of the successful and continuing interplay between organism and environment. Taoism, based on the teachings of the *Nei Ching*, advises us to seek a balance between the yang and the yin. The yang represents the sun and sky, the masculine and active, while the yin embodies darkness, the earth and moon, femininity and softness.

Much that is written on sunlight today is all yin and no yang. Contemporary reports focus on one narrow aspect such as the skin or the eyes, reduce sunlight to ultraviolet radiation, and advise us to avoid it whenever possible.

Modern scientific methods dictate a reduction of the organism to its component parts, and researchers will not rest until the person has been reduced to systems and organs, tissues and cells, and, finally, to molecules. Physicians become endocrinologists, dermatologists, otolaryngologists, and ophthalmologists. We go to the doctor expecting a diagnosis and cure of a particular ailment.

Most of us are quite willing to visit the doctor when we are assaulted by acute symptoms of disease. Few of us would attempt to treat our own appendicitis or heart attack. But seeking optimum health in this setting is another matter. Health refers to the efficient functioning of the person, the whole organism, in a given environment. In maintaining health, we are our own physicians, for we are the ones best suited to guide our actions, our habits, and our lives.

To answer the question, "How much sun is good for me?" it behooves us to note sunlight's effect on the whole organism. Only by viewing both the pros and cons in this holistic manner can our decisions be sound. Whether trying to improve or maintain health, we need to achieve the proper balance between work and rest and take the air, water, nutrients, and sunlight from the environment as needed.

Balance and moderation are the secrets of good health. The person who learns of the salutary effects of sunlight on blood and determines to remain bronzed year-round, and the person who learns that sunlight causes skin cancer and remains indoors are both violating this basic precept. Eating need not become gluttony, a glass of wine with dinner need not become alcoholism, and RDAs need not become megadoses. The same can be said for sunlight.

1

THERE GOES THE SUN

Why hast thou left the light of sun, thou poor one?
—Homer, *The Odyssey*

EARLY cultures everywhere recognized the importance of sunlight and many revered or deified the sun. Even we Americans have in recent years become a nation of sun worshippers. If, however, the contemporary heliophile were to be asked what exactly he or she found worthy of reverence in sunlight, the answer would be far different from the one given by the Mayas, Aztecs, and others who built temples to the sun and worshipped it as the creator of all life. And the answer would differ immensely from that given by people early in this century who appealed to sunlight to cure everything from tuberculosis to surgical and battle-inflicted wounds.

Today, most of us do not attribute many health benefits to the rays of the sun. We know that they will darken us, and the healthy tan is sought by many. In addition, people report a feeling of well-being when the sun is shining, but they usually assume that this is due to something psychological, and few laypersons or scientists would venture that this emotional tone is related to physiological effects of sunlight on the body.

Medical scientists, in fact, have been busy extending the list of health problems caused by solar radiation. While earlier physicians extolled the benefits of sunlight, it is presently being called the greatest of all carcinogens by dermatologists because of its role in causing skin cancer, the most widespread of all cancers. In medical circles, sunlight has passed from benefactor to bane, from friend to foe. How did this happen? To uncover the answer we must trace the course of thought about sunlight from monotheistic sun worship to medical opinion old and new.

SUN WORSHIP AROUND THE WORLD

Few cultures throughout recorded history have failed to see the connection between the sun, our day star, and all of life. The ability of plants to store the sun's energy, to capture and keep sunshine, as it were, was empirically understood long before the mechanism of photosynthesis was specified. While we know that sunshine allows plants to form carbohydrates from water and carbon dioxide, the ancients knew the sun was "Our onlie begetter," that we were born of sun and are sustained by it.

The worship of the sun is not difficult to understand. Though Christians commonly denounced such worship as paganism, the renowned Jesuit astronomer, Fr. Angelo Sechhi, wrote in the late nineteenth century: "Several peoples of antiquity worshipped the Sun, an error perhaps less degrading than many another since the star is the most perfect image of the Divine, the instrument whereby the Creator communicates almost all his blessings in the physical sphere."

In primitive cultures, the sun's intimate tie to fertility was readily understood. The connection was not limited to the sun's effects on plants and earth; the belief that humans could be impregnated by sunlight was also common. The Indians of Guacheta, Colombia, held that the sun was capable of impregnating virginal maidens, and legend had it that one of their chiefs desired that his daughter conceive a child in such a miraculous fashion. Each day she climbed an east-facing hill to receive the first rays of morning. Conception was achieved and in nine months the daughter gave birth to an emerald, which developed into a child, a true son of the sun.

Each culture had its own creation myth, an account of the birth of the sun itself. In the Kalahari Desert of Namibia, the San once believed that the sun was originally an earthbound mortal, an old man who created sunlight by raising his hands and radiating it from his armpits. But the influence of the sun was only local, limited to the immediate vicinity of the old man. The people wanted to extend the sun's domain so that the whole earth might be illuminated. This was achieved when, in a concerted effort, they seized him by the arms and legs and hurled him skyward. "Stand fast," they cried. "Thou must stand fast, then move through the sky while thou art hot." The effort was successful; the heavenly sun was thus created.

Sun Worship in Ancient Egypt

Many not-so-primitive cultures also deified the sun. Ancient Egypt had a well-developed sun religion, and in the fourteenth century B.C., Pharaoh Akhenaton expressed his praise in this eloquent hymn: "Beautiful is your rising in the horizon of heaven, living Sun, you who were first at the beginning of things. You shine in the horizon of the East, you fill every land with your beauty. You are beautiful and great and shining. Your rays embrace the lands to the limits of all that you have made. You are far, but your rays are on the Earth. . . . The beings of Earth are formed under your hand as you have wanted them. You rise and they live. Their eyes look at your beauty until you set and all work comes to a stop as you set in the West."

Author of the above hymn and ruler of Egypt's XVIII dynasty, Akhenaton, in the midst of his reign, changed his name from the royal patronym, Amenhotep, in honor of the sun god Aton. With his wife Nefretete, he undertook a total reform of Egyptian political, social, and artistic life, inaugurating a monotheism centered around Aton.

Ancient Greek and Roman Sun Worship

The sun played a major role in Greek and Roman religion as well. Apollo, Olympian god of light, was worshipped in Delphi and Delos. But he was a multifaceted god and was also the deity of music and poetry as well as serving to announce to humans the decisions of the immortals. Significantly, Apollo was also the god of healing. While Apollo represented the spiritual aspects of the sun, Helios had charge of the material aspects. Helios was worshipped as the national god in Rhodes, and the Colossus—an immense bronze statue and one of the seven wonders of the ancient world—was erected in Rhodes's harbor in his honor. He was called Sol by the Romans.

The early Greeks and Romans were far more sophisticated in the science of astronomy than in the more prosaic, earthly sciences of botany and zoology. The very people who could name every celestial constellation were at a loss to name any but the most common plants and animals with which they shared the land. And the Greeks, who believed that bees sprang spontaneously from the rotting flesh of dead animals and that swallows flew underwater to hibernate, could accurately predict a solar eclipse.

Eclipses of the Sun

Fear, among other things, is a stimulus to religious practice. Because the sun's importance for life was recognized, its loss was a terrifying prospect. What if it failed to rise one morning? What if, after the winter solstice, it didn't begin its march ever higher in the southern sky and bring the warmth of spring? Even worse, what if it just disappeared from the sky in the middle of the day?

Solar eclipses, the daytime disappearances of the sun, are rare phenomena and have been accompanied by great fear and magical attempts to persuade the sun's return. Every eighteen years, eleven and one-third days, the moon interposes between sun and earth, changing day to night over a thirty-mile diameter, and in prescientific cultures eclipses were taken to mean that the sun had been kidnapped by demons or evil people and that a ransom was necessary to secure its return. Exhortation and sacrifice were employed, anything to encourage the sun god to again make his presence felt. Using homeopathic magic, the Ojibways of south central Canada shot fire arrows into the sky to rekindle the lost luminescence of their deity.

Stonehenge and Western European Sun Worship

We know less about the earliest sun worship in western Europe than we do about the Egyptians, Greeks, and Romans. The megalithic cathedral at Stonehenge, erected on the Salisbury Plain of England in about 2800 B.C., remains something of a mystery. Written records are absent, but Stonehenge's function as an astronomical site is agreed upon. Perhaps it was used as a solar observatory, enabling ancient astronomers to determine the summer solstice. As an observatory, however, its alignments are rather crude and it more likely served as a symbol of reverence for the sun.

Celebrations of solstices and sun worship in general were active in Europe until the Christian era. The last functioning sun-cult center in Europe, a Mithraic shrine in Roman London, fell into decline about sixteen centuries ago. Concurrent with the spread of Christian theology, which admitted but one God, the pagan worship of the sun and other natural phenomena was expressly forbidden, and it virtually disappeared.

And God Said, "Let there be light"

The Bible's first divine fiat, the Hymn to Creation, expresses the master theme of Christianity. The hymn asserts that God existed prior to His creations and is the ultimate cause of them all—day, night, the firmament of heaven and its celestial bodies, as well as the earth and all that grows on it. The immanent gods of the pre-Christian pantheons were denounced as idols and their worship forbidden. The fecundity of the earth, while involving sun and rain, was ultimately attributed to the hand of Yahweh. Nature, the early Christians believed, cooperated with their deity in bringing about the bounty of the land and skies, but only in the sense that the farmer cooperates with the forces of nature to make the land productive.

The Book of Genesis tells us that light came into being on the first day of creation, while sun, moon, and other heavenly bodies were not given until the fourth day. This chronology has given theologians, from Augustine to Aquinas to the present, considerable difficulty. Could there be light without sun? Critics as far back as fifteen hundred years ago have had fun with this anomaly of physics. St. Thomas Aquinas, in his *Summa Theologica*, replied to the critics that the light made on the first day was spiritual, that made on the fourth, corporeal. Thus, the principle of light was given before the actual celestial bodies were formed. Or, as Aquinas says, sun, moon, and stars were but adornments, gracing the heavens by their movements much as clothing graces the human form.

Fruits, herbs, and grasses likewise precede sunlight in the Hymn to Creation. St. John Chrysostom, patriarch of Constantinople in the fourth century A.D., explained that this incongruity was intentional and served to guard the people against their penchant for idolatry. The sun, so frequently a god to the pagans, was mentioned far down the list of creations so there would be no mistaking which came first, God or sun, "lest perhaps lifting thy eyes to heaven, thou see the sun, and the moon and all the stars of heaven, and being deceived by error, thou adore and serve them, which the Lord thy God created for the service of all nations" (Deuteronomy 4:19).

"Be not afraid," Christians were told, "of the signs of heaven, which the heathens fear" (Jeremiah 10:2). There was nothing to worship—or fear—except He who had created them all.

As organized sun worship was disappearing from the post-Christian European scene, it was still flourishing in the Americas. Mayas, Aztecs, and Incas, unaware of Christian theology until the sixteenth century, developed sophisticated forms of sun worship.

Sun Worship in the Americas
Strangely, we have known of sun worship in our own hemisphere for a relatively brief time. While extensive accounts of Egyptian, Greek, and Roman practices have been available for centuries, knowledge of nearby cultures has been lacking.

The Mayas of eastern Mexico and Central America represent what was perhaps the earliest elaborate development of sun worship anywhere. We know that their culture dates back to the seventeenth century B.C., but the later cities and ceremonial complexes of the Classic Era yield the most evidence of their lives. Between A.D. 300 and 900, the Mayas built the imposing temples at Palenque, Bonampak, and Tikal. The priests who attended these sacred sites were accomplished astronomers and their observations have proven remarkably accurate.

The sun was a calendar as well as a deity for the Mayas, and attention to its position in the sky indicated times for planting and the beginning of the rainy season. The murals at Bonampak, in Chiapas, Mexico, reveal the practice of human sacrifice for the benefit of the sun god. The Aztecs of central Mexico similarly practiced extensive human sacrifice, tearing the hearts from thousands of victims each year so that the sun god would be strengthened in his daily journey.

When the Spaniards conquered Mexico in the sixteenth century, they were repelled by the rituals performed for the solar deities. The human sacrifices—sometimes thousands at a time when a major temple was dedicated—were performed by priests with blood- and filth-caked hair who never bathed (and who, by necessity, burned much copal incense). Wars were fought not for conquest but to replenish the supply of sacrificial victims.

Though human sacrifice may be difficult for us to understand, for the Mayas and Aztecs, the sun and rain were the givers of life. To propitiate the sun gods Quetzalcoatl (Aztec) and Kukulcan (Maya), they reasoned that they should pay in the same currency—life.

The Peruvian Incas also had a pantheon of sun gods, and the rulers were considered children of the sun, thereby deriving their divine right to rule.

Among the Pueblo Indians of the American Southwest, the sun is still a deity. And until recently, sun-watchers were used to note the exact point on the horizon at which the sun rose. Careful observation revealed that on each day the sun rose at a different point, and by keeping track of the position of the sun's first appearance behind the monuments, mesas, and mountains, the Hopis derived a natural calendar.

The last sun-watcher died in 1953, and the more familiar calendars now fulfill their function for the Hopis.

Celebrations of the Sun

It is not difficult to guess why the ancients were as concerned with the sun as they were. They were well aware of the relationship between the height of the sun and plant growth. They certainly noticed, in temperate climates, what freezing did to the crops and wild plants on which they fed. Thus, it was important to know when the sun would return to a height sufficient to sustain their spring gardens and the bounty that nature allowed them to gather in the wilderness.

The winter solstice, which occurs in late December, was celebrated around the world. In Christian countries it is known as Christmas, but long before the birth of Christ, the observance was a celebration of the celestial sun, not the Son of God.

By late December the sun has completed its drift down the southern sky (in the northern hemisphere) and begins its daily march to the north, to a higher position in the sky. The days grow longer and, in a month's time, warmer.

Three months later, near the twenty-first of March, the spring equinox arrives. Day and night are equal in length and, depending on latitude, the time for planting is near.

In the American Southwest, Indians built immense structures of adobe with carefully aligned apertures through which the spring sun could pass only at the equinox. Why did they care? An obvious answer is that they needed to know when to plant their cotton, corn, beans, and squash. And most important, they needed to know when they could expect the summer rains that would moisten the thirsty desert and save their plants from desiccation.

But why undertake the monumental task of constructing such gargantuan observatories? Two sticks and a rock would indicate an equinox or a solstice just as accurately. Besides, it is difficult not to notice that the sun is always moving north or south during the year. By noting exactly where the sun rises each morning, by recording its precise position over the landscape, a very accurate calendar can be constructed. The Hopis did just this.

The early "astronomical sites" were much more than that.

In all probability they were more like shrines or temples, or, even better, magical places. The Indian holy men not only observed and recorded solstice and equinox, they also performed rituals and supplications. The anthropologist Bronislaw Malinowski, in his classic essay *Magic, Science, and Religion*, noted that primitive people not only worshipped nature— sun, moon, and rain—they were also fearful of it. What if spring, planting, summer, and harvest failed to come this time? Everyone was aware that the rains sometimes didn't come. This fear of natural catastrophe was channeled into magical attempts to persuade the gods to deliver what was needed and to withhold their tempests. Rituals were performed within the sight of the gods so that they would be favorably disposed. The "astronomical sites" were actually places of power; the effort that went into their construction was an offering.

In our homes, comforted by central heat and glowing lights, we needn't and don't perform ritual supplications to the sun god. The seasons will come and go in obedience to natural law, not the caprices of gods. Science has quieted our fears and dulled our senses. Solstices and equinoxes pass unnoticed. But like the ancients, we are still at the mercy of the sun and its seasons as we would soon find out if spring failed to arrive one year. We would see that we are as closely linked to nature as the weed that dies in October.

The sun's important role in human history is indisputable, but in the nineteenth century, a new way of looking at such phenomena as sunlight emerged. If the ancients knew the benefits of sunlight, it was in a general way, and the mystical and physical were freely mixed. No attempt was made to isolate nature from religion or either of these from the habits of daily life. The thoughts, prayers, and labors of the people were permeated with nature.

But with the scientific revolution (and its early Greek and Roman antecedents) came a new way of thinking. The empirical, or observable, was prized above the mystical and the intangible. Scientific principles demanded that phenomena be isolated and controlled. The sun was now seen as the cause of numerous specific and observable reactions both in the human organism and in all the earth's inhabitants. The nature of these reactions and their mechanisms were addressed one by one, and some fascinating things were discovered.

THE HEALTH SCIENCES DISCOVER THE SUN

While the ancient Greeks and Romans deified the sun, their learned men also suggested specific benefits of sunshine for human health. The Greek historian Herodotus attributed many health-sustaining and curative powers to sunlight. As he traveled in foreign lands and inspected the remnants of battle, a curious phenomenon captured his attention. In Egypt's northern desert, where the Persians and the Egyptians had fought a recent battle, the natives pointed out piles of human bones. "The bones of the slain," Herodotus observed, "lie scattered upon the field in two lots, those of the Persians in one place by themselves, as the bodies lay at first—those of the Egyptians in another place apart from them. If, then, you strike the Persian skulls, even with a pebble, they are so weak that you break a hole in them; but the Egyptian skulls are so strong that you may smite them with a stone and you will scarcely break them in. They gave me the following reason for this difference, which seemed to me likely enough: The Egyptians from early childhood have the head shaved, and so by the action of the sun the skull becomes thick and hard. . . . The Persians, on the other hand, have feeble skulls, because they keep themselves shaded from the first, wearing turbans on their heads."

Though this hypothesis may sound faintly fanciful, we will soon see the influence that sunshine plays in the formation of strong bone.

From the early Romans, Herodotus learned the utility of sunshine in curing the infirm and in restoring weak muscles, and he recommended that "one should take care that in winter, spring, and autumn the sun should have direct access to the sick person."

Hippocrates and other Greek physicians prescribed sunbaths for wasted muscles, obesity, and for slow-healing wounds.

And Pliny attributed the success of Roman military campaigns to the therapeutic effects of sunbathing on the Roman people, among whom the practice was widespread.

Greeks and Romans both built special structures for the taking of their sunbaths. While some simply lay down on the terraces of their houses or on the bare sand, excavations near the temple of Aesculapius have revealed a long open-air gallery connected to a sick ward. Here convalescents received supervised solar therapy, or heliosis, as the Greeks called it.

The Romans built solariums, often terraces on the roofs of houses where sun could be taken in peace and privacy. And foreshadowing a development of our own century, they treated their tubercular patients with a combination of sunlight and mineral-salt spas.

The advent of the Christian era heralded a long, dark spell not only for sun worshippers. Early applied work on the therapeutic use of sunshine was also stultified. The wisdom of the Greeks and Romans was lost, and through the middle of the eighteenth century, little is to be found on heliotherapy (healing with sunlight) in the Christian world.

But two events in the latter part of the eighteenth century were to change this. Jean Jacques Rousseau, French philosopher, argued persuasively that the natural life was the good life, that a feeling and regard for nature was superior to the perverting artifices of church and state. This back-to-nature movement, together with increasing sophistication in the field of optics, ushered in a new appreciation for the powers of the sun.

Folk wisdom has always had something to say about the influence of sunlight on health. An old Italian proverb maintains that "when the sun doesn't enter, the doctor does." But during the late eighteenth and early nineteenth centuries, physicians began to pin down some of the specific curative attributes of solar radiation.

In 1774 physicians in France began treating open ulcers of the leg with sunlight, attributing their success to solar heat. Soon after came reports of cancer cures achieved by augmenting the sun's power with lenses.

J. F. Cauvin, on the basis of his careful observations, claimed that sunlight could cure muscle weakness, rickets, scurvy, rheumatism, dropsy (the accumulation of fluid), scrofula (tuberculosis of the cervical lymph glands), and paralysis. Florence Nightingale similarly recognized the curative powers of the sun and insisted that sunlight be allowed to stream into the hospitals that she established during the Crimean War. Important experimental support for some of these observations was obtained in 1877 when it was shown that sunlight could kill bacteria.

Climatotherapy and Western Migration

At about this time, an innovative branch of medicine was established in the United States. Dr. Charles Denison recommended that physicians inaugurate the formal study of the effects of climate on disease, that they study climatotherapy. Denison's recommendation came at a time when the list of intractable diseases was lengthy—medical science of the day could offer no cures for common afflictions such as tuberculosis (then called consumption), malaria, or dysentery. Contemporary theories emphasized qualities of the atmosphere as the cause of diseases of unknown origin. Many maladies were thought to originate from putrid vegetation and meat (which breeds maggots), animal exhalations, and warm, moist places in general.

Disease was more prevalent in some places than in others. In the nineteenth century, the Mississippi Valley was a notoriously unhealthy place to live, with tuberculosis, dysentery, and malaria being widespread. As climatotherapy grew in popularity, doctors who could offer no definitive cures for these persistent ailments increasingly offered this piece of advice: "Go west." The change of climate, living in a place with abundant sunshine and dry air, presented the hope of effecting cures the physician could not. The lower Southwest, in particular, was acclaimed for its dry, elevated, sun-drenched deserts and mountains. As Billy Jones notes in *Health-seekers in the Southwest, 1817–1900*, such places were considered to be nature's sanitoriums. The movement of the population westward was not only a result of people's search for economic fortune or because the West offered a haven for misfits who did not enjoy a close proximity to the law. It also received significant impetus from health seekers.

Early adventurers in the West seldom put their thoughts into print without commenting on the healthful aspects of the sunny, dry climate. In writing of his mapmaking expedition from Pittsburgh to the Rocky Mountains, Major Stephen Long noted that "The Indians' catalog of diseases and morbid afflictions is infinitely less extensive than that of civilized man. No case of phthisis [TB of the lungs] or jaundice fell under our observation. . . . They are rarely afflicted with dysentery and are never known to be subject to sunstroke, although they travel for days and even weeks over the unsheltered prairies."

Edward Fitzgerald Beals, who laid out a wagon road in Arizona, wrote: "In the journey of the year during which I have been engaged in this work, I have not lost a man nor was there the slightest case of sickness in camp; the medicine chest proved only an encumbrance. My surgeon having left me at the commencement of the journey, I did not employ, nor did I have need of one on the entire road."

And easterners listened. Toward the end of the nineteenth century, one source estimated that fully one-third of Colorado's population had moved there for purposes of health. California, Arizona, New Mexico, and Texas also attracted health seekers.

Tuberculars in particular sought the benefits of climate and were readily welcomed by whatever version of the Chamber of Commerce western cities had at the time. Hotels were built expressly for invalids. For those less able to afford such sumptuous accommodations, tent cities sprang up on the outskirts of town and in Tucson an unbroken row of tents stretched from the city limits to the foothills of the Santa Catalina Mountains several miles distant.

At about the turn of the century, things changed. Researchers discovered that tuberculosis was transmitted by bacilli and that, furthermore, it was contagious. Promoters who had previously opened their desert and mountain towns to tuberculars now made it known that the "lungers" were no longer welcome. And residents who had lived without concern in the proximity of TB camps began to keep their distance. Tuberculars were forced to live as nomadic desert rats, out of tents and wagons far removed from others.

People still seek out the sun for their health. Arthritics and asthmatics continue to move to the Southwest for its climate. And psoriasis sufferers visit clinics at the Dead Sea. But climatotherapy as a formal discipline went into a serious decline as a new science, bacteriology, took over. Medical researchers became far more concerned with the identification of invasive microscopic agents that entered the organism to produce illness. But if the study of which geographical location could best cure disease was finished, the investigation of what sunlight could do for the body was not.

The Healing Sun

In 1877 scientists demonstrated that sunlight kills bacteria. But a more significant finding was made fifteen years later with the discovery that it was the short, invisible wavelengths—the UVR*—that had the most pronounced bactericidal effect. These shorter UV wavelengths, it was learned, have a greater action on the surface of the body than do rays of visible light, and researchers were eager to discover what other applications they might have.

Niels Finsen, who has been called the father of heliotherapy and phototherapy (treatment with artificial light), demonstrated conclusively that it was the UV wavelengths that produced sunburning and tanning of the skin. Protecting certain parts of his arm with glass plates while leaving other areas bare, Finsen exposed his arm to UVR. He knew that the shorter UV wavelengths were blocked by plain glass. As expected, only the fully exposed portions of the arm showed the darkening reaction. Finsen reasoned that if the UV radiation was

---

*abbreviations: UVR = ultraviolet radiation; UV = ultraviolet

responsible for inducing redness and pigmentation in the skin, whereas visible light was not, perhaps there were other effects of this radiation still to be discovered.

In Denmark, Finsen's home, tuberculosis of the skin (lupus vulgaris) was widespread. Those afflicted developed ulcers on the face and neck that, even if cured, left ugly scars. For long periods of the year in these northern latitudes, the sun is weak or absent—perhaps, thought Finsen, this contributed to the disease. He showed that by concentrating sunlight with a quartz lens (which allows UV to pass through), dramatic cures of the ulcers were effected. He later turned to the use of a type of artificial light rich in UVR, feeling that this source was more reliable and controllable than sunlight. In 1903, the year before his death, Niels Finsen was awarded the Nobel Prize in Physiology and Medicine for his efforts.

Important discoveries were being made elsewhere at the same time. Paradoxically, sunlight was used to treat sunburn. And artificial light was concentrated and found to lower blood pressure. Patients treated with UVR every five days showed a decrease in blood pressure of between 8 and 10 percent. The effect lasted for months. Those suffering from angina pectoris reported marked relief from pain and freedom from new attacks.

The Swiss physician Oscar Bernhard, after observing the effectivenesss of mountain sun in curing meat, was encouraged to try sunlight on human wounds. He wrote: "On the night of the second of February, 1902, an Italian with seven stab wounds was brought into the hospital. Amongst other wounds there was a perforating chest and two perforating abdominal wounds with injury to the liver and spleen. Eight days later the wound opened along its entire length. . . . It was impossible to bring the edges of the wound together. . . .

"As I came into the hospital one beautiful, clear morning the sun shone warm through the open windows and the fresh, exhilarating air filled the entire ward. The thought suddenly occurred to me to expose this large wound to sun and air. I remembered that the mountain peasants always hung their meat in the dry air and thus cured it in the sun and I decided to use this antiseptic and drying action of the sun and air on this refractory wound. . . . Within an hour and a half there was a marked improvement and the appearance of the wound had entirely changed." The wound soon healed normally.

In 1919 there came a discovery that would have an even wider application. Sunlight, with its UVR, could cure rickets, a condition in which calcium is not absorbed normally and which is marked by a bending and distortion of the bones. Sunlight, as we now know, promotes the synthesis of vitamin D in the skin, facilitating the absorption of calcium by the gut and preventing rickets. The early Egyptians apparently understood that sunlight cures rickets—we have stone plates showing Akhenaton and his wife Nefretete giving sunbaths to their rachitic children.

Just a century ago, rickets was a major health problem, especially within cities lying in northern latitudes. London, in particular, with its pall of coal smoke and its narrow streets shadowed by tall buildings, had more cases of rickets than any city in the world. The discovery that small amounts of sunshine could cure and prevent this disease has been one of modern medicine's most important breakthroughs.

Many other claims have been made for the efficacy of light in treating and preventing disease. As already noted, the climate cure was prescribed for TB of the lungs and other widespread infections for which there were, at the time, no medicines. And a variety of skin disorders have been treated with light. But by the turn of the century, light, both natural and artificial, was being used in a controlled fashion by trained professionals. Instead of being told to get more sun or, more succinctly, to move, patients were now given detailed sunning prescriptions to follow or were guided to rooms with complex equipment. Control was the key.

Controlled Treatment: Heliotherapy Comes of Age

In 1903 the Swiss physician Auguste Rollier opened his world-famous clinic at Leysin, high in the Alps. Initially, his clients were people suffering skin maladies such as skin tuberculosis, complications following surgery, and all manner of wounds.

Rollier's "solar dressing" for wounds was, he thought, far superior to ordinary dressing, but only if the exposure to sun was carefully calculated. His scheme of "progressive insolation" called for baring only the feet on the first day for a period of five minutes. The rest of the body was clothed in white linen, with a white hat over the patient's head, a screen over the face, and smoked glass shielding the eyes. On the second day, the feet were exposed for ten minutes and the lower legs for five. On each successive day, an additional part was exposed—thighs, abdomen, and chest until in two weeks, the entire body was receiving sunshine for between one and three hours a day. Patients' precise prescriptions for sun dosage depended on a number of factors such as season, time of day, and humidity. And the individual's age and sensitivity to sunlight were taken into consideration.

Even serious burns from fire were treated in Rollier's clinic, but in this case sunlight exposure was received only once, for a very few minutes, and through gauze. The remainder of the healing was accomplished by exposure of the wound to mountain air.

But it was the treatment of tuberculosis that won Rollier his greatest acclaim.

Rollier must be given credit for the popularization of heliotherapy in our own century. At his Leysin clinic, he began treating sufferers of the deforming tuberculoses of the bones and joints. To his clinic, which eventually grew in size to over 1,200 beds, twisted tuberculars came in large numbers to receive the miraculous cure. Using the system of progressive insolation, patients were exposed to increasing amounts of sunlight until the entire body was bronzed.

As early as 1892, in fact, Dr. Antoine Poncet was making use of heliotherapy to treat similar diseases, but it fell to Rollier to popularize and propagate the technique throughout the world. His patients were considered hopeless by the medical world, and under traditional therapy their prognoses would have been most somber. But Rollier was able to improve the condition of 90 percent of his patients, achieving outright cures in 78 percent of the cases. In February 1912, Dr. Rollier astounded the Society of Physicians at Leysin by displaying numerous photographs and X-rays of patients with tubercular limbs, nearly all appearing to call for amputation. But photographs of the same patients taken after many months of heliotherapy revealed an incredible improvement in the most warped bodies.

As noted, Rollier did not limit heliotherapy to the afflicted region of the body. The whole body (with a small area excepted for purposes of modesty) was exposed. He believed that tuberculosis was an infectious disease, one that indicated that the entire substrate of the organism required rehabilitation. By treating the whole body, the individual's general resistance was strengthened and could heal the local site of the affliction. In addition to repairing the tubercular bone or joint, the whole organism was transformed.

Word of Rollier's success spread, and before long similar clinics were established on the shores of the Mediterranean Sea and Atlantic Ocean as well as at Alton, England.

The early developments in heliotherapy took place largely in Denmark, France, and Switzerland. The technique was applied enthusiastically in Germany and Russia as well. But very little was done in the United States, and with the exception of climatotherapy's brief flurry, America appeared to have little interest in the healing powers of sunlight until 1914 when heliotherapy hit our shores with a vengeance.

Heliotherapy in America

Prior to 1914, very little was heard about the benefits of sunlight by people in this country. Scientific journals had been publishing articles on the germicidal effects of the sun and one could find a few how-to articles on building a sun parlor on to the house. But the most frequent references to sunlight in the contemporary literature were negative, the most popular issue being how to avoid sunstroke. But in February 1914, *Literary Digest* broke the news to the general public—sunlight had miraculous curative powers. Articles with titles such as "Miracles Wrought by Sunshine," "Sun Cure," and "Sun as Surgeon" began appearing in the popular press with regularity. During the same year, a sunbathing fad swept California, and Golden Staters could be found lying in the sun trying to treat every disease imaginable.

But the medical profession was also showing interest in applying the controlled, supervised techniques that had been so successful for Rollier. At the Sea Breeze Hospital at Coney Island, New York, tuberculoses of the bones, joints, and glands were treated with heliotherapy, and New York City made plans for a thousand-bed hospital for the same purpose on the beach at Rockaway Point. Amazingly, Sea Breeze Hospital was successfully treating Pott's disease, a tuberculosis of the spine that causes a pronounced hunchback due to a sharp forward angulation of the backbone. At Sea Breeze, Pott's patients were enveloped in a plaster cast that covered the entire torso and neck, but the chest portion of the cast was cut away to allow sun exposure. Before-and-after photographs reveal the striking improvements in spinal alignment after several months of heliotherapy.

In Perrysburg, New York, the J. N. Adam Memorial Hospital was opened in 1914 to treat Buffalo's tubercular children. Youngsters roamed the pastoral grounds clad in nothing but white hats and loincloths reminiscent of diapers. The expected benefits of this therapy were elucidated in the October 1914 issue of the *Survey:* "This is a glorious stimulus for nature study to the waif who has heretofore known only the city streets or slums. . . . Here they learn of nature's abundance, feel her restoring powers, respond to her touch and in their primitive costume come back to full health, for even the cold and snow will have no terror for the child of the Rollier treatment; for he is already resistant to them. They will but render him more hardy and send him back to civilization not merely well but vigorous and rugged." In the United States, Rousseau's back-to-nature call was to be heeded more fully only by the nude culturalists.

During this era, sunlight was being recommended as a specific antidote to old age. In addition to relieving depression and improving the aging person's ability to work, Arnold Lorand, M.D., suggested that sunlight relieves the kidneys of part of their burden by stimulating perspiration and the elimination of toxins through the skin. Sunlight, noted Lorand, also dilates the blood vessels and sends blood to the periphery of the body, thus improving circulation. "Against old age sunlight should be regarded as an excellent protection." Both youth and life could, with sunshine's aid, be prolonged.

The sunlight movement was gathering momentum. During the years of World War I, while innovative research did not progress much, the application of solar therapy proliferated and was used to treat the war wounded, most notably at the American Hospital in France, with excellent results.

In April 1921, Dr. E. Roux reported to the French Academy of Sciences on the wide assortment of ailments that had, to date, responded to sunlight therapy. Numbered among the successes were "wounds, including varicose ulcers, burns, surgical wounds, etc., as well as bacillary fisculas, tuberculosis of the ganglia, lupus, chronic arthritis of tuberculous or rheumatic nature, sciatic neuralgia, and other forms of neuralgia. Still another group of invalids treated by means of heliotherapy includes a certain number of persons suffering from chronic tuberculosis of the lungs."

In the years following 1914, a general interest grew in sunlight and health. But it was not only the ailing who sought the sun—others believed that sunshine would help them maintain their health or make good health better. People were now especially eager to let sunlight stream into their houses, and the home darkened by thick shades or drawn blinds was seen as the domicile of bacteria and ill health. Those building their homes were advised not to square the structure with the cardinal compass points, since this insured that one side of the house would never see sunlight. Instead, if homes were oriented at 45 degrees from the cardinal points, all walls would receive sunshine during at least part of the year. (The wall facing NE or NW would catch the early morning or late after-noon sun, respectively, in the summertime when the sun reaches its northernmost position on the horizon.)

Since the early part of the nineteenth century, cities have been condemned for depriving their inhabitants of sunlight. With their multistoried structures and narrow streets, with limited space for outdoor recreation for young and old alike, cities were recognized as havens of disease. In 1827 an ob-server noted: "If we wish to etiolate men and women, we have only to congregate them in cities where they are pretty securely kept out of the sun and where they become white, tender, and watery as the finest celery."

Until 1851 residents of London and other cities were addi-tionally deprived of sunlight. In England, as in many other countries, a window tax was levied and the poor were forced to choose between food and sunlight.

In the United States, architects began planning "sunlight cities." In a periodical entitled *The American City* (Septem-ber 1917), Herbert Swan and George Tuttle decried the street and building plans of contemporary metropolises. In Manhat-tan, they noted, the majority of factories, offices, and apart-ments received no sunlight at all in the winter. Skyscrapers could cast shadows more than 1,500 feet in length on the shortest day of the year. At noontime on December 21, for example, Swan and Tuttle calculated that the Equitable Building cast a shadow one-fifth of a mile in length, com-pletely enveloping an area of seven and a half acres.

The envisioned "sunlight cities" would have wide streets and large open spaces, a design of greater necessity in north-ern latitudes. Tall buildings were to be built on streets run-ning north-south, and they recommended that apartments on the south side of an east-west street have windows in the back. With careful planning, a sunlight city in the latitude of Win-nipeg would give each dwelling more sunshine than a non-sunlight plan in Key West.

The sunlight pendulum had never swung as far in the United States as it did during this period. Practically everything that was being written was sanguine. Skin cancer was not mentioned. Some current myths—that glare could result in permanent eye damage, for example—were addressed and dispelled. One physician, Dr. F. Robbins, writing in the *Medical Record*, suggested none too subtly that there was probably something wrong with people who didn't like sunshine. He noted that those who complained of bright sunlight were generally neurasthenic and hysterical individuals—or worse, possibly drug addicts. "Shrinking from bright daylight or sunshine," Robbins wrote, "is a very common sign of neurasthenia. The victims of drug habits, especially morphine, often insist upon the most rigid exclusion of light from their shadow realm."

It should be emphasized that practically no one, no matter how enthusiastic he was about sunlight, advocated unrestrained exposure to it. Most prescribed very careful regimens of exposure. One writer, while attributing Jack Dempsey's victory over Carpentier to Dempsey's "elastic tanned skin which Carpentier's blows did not harm," specified some rather unusual precautions to be observed while achieving one's tan. The sun worshipper "in the early summer, should wear green goggles and sit under a red parasol in a red silk bathing suit"(!)

Sunbathing and the Healthy Tan

Without a doubt, the term "healthy tan" was coined during this period. Pallid skin was equated with ill health. Furthermore, though physicians who used heliotherapy claimed no intrinsic benefits for pigmented skin, they used the intensity of the patient's tan as one index of how well therapy was progressing. But with a broad sweep of logic, many people took this index of improvement among the *sick* to mean that the darker *everyone* was the better. The plight of blonds was bemoaned at length, since they could never become as dark as their more fortunately endowed dark-haired friends. Nudists carried the argument to its logical extreme by exposing everything all the time.

While the Greeks and Romans of two thousand years past had solariums in their homes so they could sunbathe, sunning oneself did not become an avocation until the present century. In the nineteenth century, the lady of leisure was known by her pale, ivory complexion. One could tell by her pallor that she was of high social standing, not needing to work in farm or field to earn her keep. With the advent of factories, however, the picture changed. Now working-class folk, sequestered behind brick walls for the whole day, readily acquired the pale look. A suntan, consequently, became the new status symbol. And in the wintertime, a tan clearly demonstrated that one had the means to travel to sunnier locales.

A letter writer from France commented on this new phenomenon: "Sunbaths were much in vogue on all the beaches of France during the summer of 1928, which was characterized by exceptional luminosity and heat. The long lines of bathers stretched along the sand for hours furnished a curious spectacle. The style demanded that the entire skin should take on the color of gingerbread. People took delight in acquiring their coat of tan; for by this evidence, they proved to their friends, on returning home, that they belonged to the leisure class that is able to afford the luxury of long vacations. But since for many this was not the case, they were forced to press into two weeks the effects of a long vacation. They therefore proceeded to expose themselves to the sun with a heroism that resulted in vast skin burns, fever, and general disorders."

While the cosmetic advantages of a tan were conferring status to some, others were pursuing health. "No tan, no benefit" became the motto.

The excessive sunbathing that normal, healthy people indulged in is analagous to the later vitamin craze in our country. After numerous vitamins were identified, synthesized, and tested, the crucial role for each was uncovered. For most, a deficiency caused recognizable and often serious symptoms. But in an unwarranted generalization, some people took these discoveries to mean that all similar symptoms were responsive to vitamin therapy. Lowered resistance? Take vitamin C. Nervous? Take some B vitamins. Poor vision? Try vitamin A. And if the Recommended Dietary Allowance did not yield the desired results, increase the dose. Megadoses were the ultimate strategy, but with the fat-soluble vitamins—A, D, and

E—some people got into trouble with overdoses. And so it was with sunshine. Since a complete lack leads to poor mineralization of bones, lowered resistance, and general weakness, among other things, unlimited amounts might, so the argument went, produce miraculous cures and improve an already healthy organism. Even criminal behavior, some proposed, would yield to sunshine therapy.

Improving on Nature

During the heyday of heliotherapy, there were many who tried to augment the sun's healing powers. For some, especially the ailing, simply sitting in the sun was not enough. New contrivances were developed to maximize solar insolation, and eventually artificial sources of irradiation were developed.

As early as 1815, L. Loebel was prescribing sunbaths for a variety of infirmities including general weakness, rheumatism, gout, and diarrhea. He invented a special box called a heliothermos with walls and roof of glass, in which patients sat, their heads protruding through a hole in the top, for their sun-heat cures.

By the 1920s a solarium had been developed at Aix-les-Bains in Savoy, France, one that revolved on a pivot and received sunlight throughout the day. The sunning booths were arranged around a physician's office so that patients in five booths could be monitored simultaneously. The structure, tipped on its axis to allow the most direct exposure, contained reflectors and lenses, thereby providing maximum concentration of the available light. Different panes were used, some admitting UV rays, others transmitting only blue, green, or red colors. In common with Rollier's clinic, the solarium at Aix-les-Bains was designed to avoid overheating of the patients. Head coverings were used and skin temperature was constantly monitored.

Color Therapy

The colored panes of glass used to filter the sunlight at Aix-les-Bains represent one of the few innovations in heliotherapy that can be traced to the United States. Color therapy arose from the work of Edwin Babbitt, a self-proclaimed magnetist and psychophysician.

In tracing the development of heliotherapy, we have to this point been concerned with the discoveries and techniques of people who practiced the traditional medical sciences. But there were others such as Babbitt, who, though often lacking medical degrees, had their own reasons for recommending the benefits of sunlight. Sometimes they called themselves harmonists or color therapists. Today we know them as holists or naturopaths. A recurrent theme in their thinking is that people should live in harmony with the natural world. Air should be uncontaminated, food unprocessed, and water pure.

Holists argue that we need the proper balance of nutrients, water, exercise, and sunlight. Babbitt, an early practitioner of this philosophy, published *The Principles of Light and Color* in 1876. Though he had no medical background, the book had a profound impact on the world. Each color was said to cause specific vibrations in the body—red was exciting and stimulating and could promote arterial blood flow and accelerate healing; blue was soothing and to be used for nervousness, sleeplessness, and the like; yellow was useful as a laxative or diuretic. Filters were placed before a light source, preferably sunlight, to achieve the desired color.

Color therapists also advised their clients—and still do today—on the color of clothing to wear, color schemes for home and office decoration, and which foods to eat. To increase vitality, red foods (tomatoes, radishes, plums, etc.) should be eaten. In the case of high blood pressure, blue foods (blueberries, grapes, etc.) are recommended.

Phototherapy

Because of Babbitt's lack of training in medicine, his theories left little mark on the medical profession. But if color therapy never captured the attention of the mainstream physician, a technological development did. Rather than modifying sunlight with filters, some preferred to replace it altogether with artificial lights.

We have seen the pattern before with Niels Finsen's work in Denmark. Finsen came to see the sun as unreliable and weak for the treatment of skin tuberculosis. Sunlight was too unpredictable and, furthermore, only a small portion of the active radiation in the UV range ever reached the earth's surface. The apparent solution was the use of artificial lights, and phototherapy was born. Finsen used a carbon arc as his light source, and soon mercury discharge lamps were developed. A patient's prescribed dose of light could now be precisely controlled and, because of the great intensity of the lamps, delivered in a much shorter period of time. Lights could also provide heat radiation to localized areas of the body without raising the whole body's temperature. And high doses could be given in the UV range.

I will discuss phototherapy more fully in Chapter Six. For now, it suffices to say that phototherapy and heliotherapy enjoyed similar acclaim during the early decades of this century. Artificial UV was being used to treat some 166 different ailments, including constipation, hemorrhoids, cirrhosis of the liver, hypertension, arteriosclerosis, varicose veins, asthma, bronchitis, pneumonia, insanity, sciatica, gout, lumbago, rheumatoid arthritis, herpes, psoriasis, vaginitis, corneal ulcer, laryngitis, ulcer of the eardrum, lupus, hay fever, sinusitis, tonsillitis, dental caries, pyorrhea, warts, and colds.

Today, however, phototherapy has fallen from the lofty position it once held, and the reasons for its decline are very similar to the ones that brought heliotherapy down.

The Pendulum Swings

By the 1920s physicians were well aware of the damaging effects of overexposure to sunlight, and the link between UVR and skin cancer in laboratory animals had already been established. Concerned doctors published articles in the popular press, but sunbathing continued to grow in popularity.

Medical journals began reporting bizarre skin reactions following sunbathing. Cases of dermatitis—with its reddening, swelling, and crusting of the skin—were noticed after sunbathing in fields. At first, meadow grass and chlorophyll were suspected, but physicians later discovered that sufferers had been in contact with plants from the furocoumarin group (containing such plants as wild parsnip, wild carrot, and giant hogweed). Contact with these plants in the presence of sunlight can cause a toxic reaction in some people. And strange, pendantlike inflammations were occurring on the faces, necks, and breasts of female sunbathers. Bergamot oil, a substance found in perfumes and other toilet articles, was finally implicated. As with the furocoumarins, this dermatitis only occurred in connection with sunlight.

Then things became more serious. A Dresden ophthalmologist, F. Schanz, reported that sunlight could harm the retina of the eye. This damage, resulting in a clouding of vision, was thought to occur as a consequence of simply being outside on a bright day. Similarly, cataracts of the lens were produced in laboratory animals when they were exposed to UVR. Thus, the same radiation that had been praised so highly by heliotherapists only a few years before was taking on the characteristics of a malefactor. What had been touted as a cure-all was now condemned as a poison.*

Solar dermatitis was not a major concern of the early heliophiles. Dermatitis could be avoided by staying away from sensitizing plants and chemicals. But retinal burns and cataracts were something else. The common reaction to the scary news about UVR was a simple one: wear sunglasses. Tans were as eagerly sought as ever, but starting in the 1920s, sunbathers began donning colored glasses. Whereas they had previously been worn only by people suffering from an abnormal sensitivity to light (who were looked upon with some pity), sunglasses now became stylish.

Other styles changed as well. As bathing suits became less

---

*Schanz's research has been discredited, as no one was able to corroborate it. Schanz himself repudiated his own work and in later years actually used UV therapy in the treatment of eye diseases. The cataracts produced in the laboratory (in fish) were later found to have been caused by using radiation with a wavelength shorter than that found on earth and thus of little concern to us. Chapters Three and Four will discuss in detail the benefits and problems of exposure to sunlight as we know them today.

modest and ever more flesh was exposed, skin cancers began occurring in novel places. For the most part previously limited to head, neck, and hands, they now appeared on the legs and on the upper chest and back areas.

Men began to develop more skin cancers in the entire chest and back regions—but not in New York. Not until 1936 was it legal to wear swimming trunks there, and in 1934 eight men were fined a dollar apiece for eschewing the shoulder-to-thigh tank suits and going topless at Coney Island. "We'll have no gorillas on our beaches," the guardians of community morality declared. The law was changed when the County Parks Commission, which rented suits to bathers, found that trunks were cheaper to purchase and maintain than were tank suits.

When women began going topless (first on the beaches of France) many cancers of the area around the pigmented areolae appeared. Our worries about the sun were expanding, and more was yet to come.

The tone of the 1930s is well represented by Dr. Charles Pabst, a dermatologist who introduced the term "heliophobe" (a person who fears the sun). Many people, Pabst explained, are very sensitive to sunlight. Those of Northern European descent, blonds, and those with blue eyes frequently burn and blister but never get the coat of tan they desire. In some cases death could result from sunburn, and among infants fatalities from even brief exposure were reported. People differ, Pabst said, in their reactions to sunlight. Sensitive persons were urged to use a sunscreen. One that had recently been developed was a substance called esculin, which was extracted from the bark of the horse chestnut tree. This counsel was to become the accepted medical position, and today we hear of the six skin types and how much sunshine each can take before there is damage. (There will be more on this subject in Chapter Two.)

The decade of the 1930s marked the beginning of the end for heliotherapy in America. Scientists invariably moderated the optimistic tones of the previous two decades when discussing the benefits of sunlight. It was widely believed that sunshine had been oversold, that it was being prescribed in-

discriminately, and that only carefully controlled dosages—
preferably under the supervision of a physician—could produce any desirable outcomes.

Researchers began to ask some tough questions. How did
heliotherapy actually work? The curing of rickets and the killing of bacteria were well established, but how did it heal burns
and wounds and improve general health? Many solariums were
in the mountains, in resortlike settings. Cures might be attributed to fresh, dry air, the mere warmth of sunlight, or to
the salutary effect of a holiday away from home and work. The
direct effect of sunlight falling on the body could be one of
the several factors effecting cures, or it might, the critics indicated, even be unrelated to health.

It is no coincidence that antibiotic drugs were developed
during this time, medicines that made infectious disease less
threatening and which successfully treated all varieties of tuberculosis and infections from wounds.

Modern Medicine

Sir Alexander Fleming's discovery of penicillin in 1928 was
to bring a halt to the vast majority of heliotherapy research,
at least in this country. Fleming, working with the staphylo-
coccus germ, noticed that the germ failed to grow around sites
where the common fungus penicillum was growing. Unfortunately, when he applied the substance to infected wounds,
results were not encouraging because, as we now know, ex-
traction procedures were flawed. But with the outbreak of
World War II, researchers mounted a concerted drive to perfect Fleming's methods, this time with success. Whereas solar
therapy was a popular technique for treating wounds and other
infections during World War I, antibiotics came into wide use
during the Second World War and completely eclipsed heliotherapy.

Soon the list of drugs used to treat infectious diseases ex-
panded, and by 1946 para-aminosalicylic acid (PAS), a chemical that had been around for years, was recognized as useful
in the treatment of tuberculosis. Isoiazid was also synthesized
and this drug along with streptomycin (an antibiotic) and PAS

made possible the cure of consumption, "the consumer of humankind."

Even earlier, in the 1920s, vitamin D had been artificially produced by irradiating a natural plant substance, ergosterol, with UVR. Rickets could now be cured without sunlight. Science, so it seemed, was rendering sunlight irrelevant.

The Sun Sets on Heliotherapy

The death knell sounded for heliotherapy in 1941 with the publication of the first popular article linking skin cancer with sunbathing. Dr. James Ewing, writing in the *Ladies' Home Journal,* introduced the public to "sunlight cancer." He dismissed sun cures categorically, with the exception of treating rickets. "Of course," he allowed, "there is some virtue in sunbathing, but the benefit does not come particularly from the direct rays of the sun. It comes from the outdoor air, fresh breezes, exercise, relaxation, improved appetite, and the general psychic sense of exhilaration which comes from the change of scene and activity."

Dr. Ewing presented several case histories of his patients who had developed skin cancers, one of whom was in danger of losing his lip to surgery. A farmer who developed a cancer on his hand and had neglected treatment for two years had to suffer amputation. "Cancer," Ewing concluded, "is too high a price to pay for the tanning habit." (Interestingly, Ewing's article followed one entitled "Vitamins Can Restore Your Health." Good-bye nature, hello pills.)

For the first time in decades, nothing favorable was said about sunlight in the popular press. The *Readers' Guide to Periodical Literature* dropped the headings "Heliotherapy" and "Sunbaths," sections that had been well represented for the previous twenty-four years.

Through the 1940s and 1950s, heliotherapy was ignored by the press, but articles on protection—the use of sunscreens and sunglasses—proliferated. At an American Medical Association convention in Dallas in 1959, the medical establishment made it official: sunlight and American sunbathing habits were responsible for the increased incidence of skin cancer, especially in the Southwest, the region people once came to for *improved* health.

In the 1960s, the most common references to sunlight were on solar energy as Americans began to understand oil's political ramifications. When sunlight was referred to in a health-related article, skin cancer, sunburn, and sunglasses ("Something New Between the Sun and You") were the most common themes.

The same concerns persisted through the 1970s, when an additional one was exposed—nude sunbathing. But it was now the social aspects, not the health issues, that were featured. Would American beaches, even isolated ones, be opened to nudity? Could the sexual revolution be extended to what people did in public? And should we invest heavily in binocular stocks? Our attention shifted as sunbathing became a social problem. Quite obviously, the health aspects of sunshine had become passé—everyone knew it was harmful.

PRESENT CONCERNS ABOUT SUNLIGHT

Evidence is accumulating that implicates sunlight in the development of melanoma, the most virulent and lethal form of skin cancer. If this is the case, then sunlight is now associated not only with the skin lesions caused by the less serious skin cancers. It may also be related to a far more dangerous type of cancer, one that can metastasize and spread throughout the body.

I have reviewed the changes in medical opinion concerning UV radiation, but until recently concern was focused only on the middle wavelengths of UVR, the UVB. Recent evidence suggests that the longer UV, the UVA, as well as infrared radiation, may also be related to skin cancer. (Research pro and con is discussed in Chapter Two.) The width of the solar spectrum we are advised to avoid is expanding.

SUNLIGHT EDUCATION

During the 1970s, public education on the effects of sunlight took a dramatic upswing. When the feasibility of building a fleet of supersonic transports (SSTs) was being debated, many people became aware of ozone as a substance that does more than choke us in our smoggy cities. Ozone absorbs much of the shorter UVR before it reaches the earth, and since the SSTs might cause a deterioration of the ozone layer, our exposure to this worrisome radiation could increase. The warning concerning the effects of Freons (used in spray cans and for refrigeration) on the ozone layer further raised public consciousness.

Also in the 1970s, popular periodicals, mainly women's and sportsmen's magazines, began doing numerous pieces on the harmful effects of sunlight on the skin. Physicians summarized the bad news and recommended hats and the use of sunscreens, among other things. This time people listened. Where they had previously purchased sun lotions that claimed to facilitate the tanning process, many consumers now turned to sunscreens, preparations that filtered out UV rays and delayed the tanning process. One could still enjoy the benefits of sunbathing but without incurring skin damage.

For men, cowboy hats and other styles of headwear became stylish. For years one of the best ways to get a laugh was to answer the question, "What do you do?" with, "I'm a hat salesman." As everyone knew, such an occupation was not too demanding and afforded unlimited amounts of leisure. Similarly, women's hats were not much in demand in recent years. But when the designers revealed their 1982 spring fashions in Paris, there they were—straw hats were back. These changes in style may or may not reflect health concerns, but the end result is that the dedicated follower of fashion will receive more protection from sunlight.

SUNSHINE REPRISE

The sun is down but not out. Until the early 1980s, it appeared evident that sunlight was going the way of cigarettes and illicit drugs—each was declared harmful, and the most sensible strategy for health educators was to inform and dissuade the public from their use. Quit smoking, turn to natural highs, and avoid sunlight. But the burial of sunlight was premature.

An article in the January 1981 issue of *Harper's Bazaar* led off with this statement: "The sun is our most erotic natural resource, scientists say, with the power to arouse the sex drive and step up performance." Though it is difficult to imagine any physiologist or solar engineer referring to the sun as "our most erotic natural resource," the article correctly reported the relationship between the pineal gland, light, and sexuality. We had another candidate, so it seemed, in our age-old search for an aphrodisiac.

Indeed there *is* a link between sunlight and sexuality. But the new sunlight-health revival does not stop there. From a variety of laboratories and clinics come reports that suggest we had better reevaluate the effects of sunlight.

Though the sun is solidly implicated in the genesis of two kinds of skin cancer, basal-cell carcinoma and squamous-cell carcinoma, recent research reveals that sunlight may also have a role in repairing the damage it has done. Light rays with wavelengths longer than the UV responsible for skin damage can repair damage to the genetic material, DNA. Called photoreactivation, this process has been known to occur in most lower organisms, but until recently evidence for its existence in mammals has been lacking.

As far back as 1940, Dr. Sigismund Peller of the Johns Hopkins School of Hygiene gathered evidence indicating that exposure to the sun may prevent cancers in areas other than the skin. Getting enough sunlight to produce a skin cancer, in fact, was related to a lower than normal incidence of more serious cancers. Peller suggested that skin cancer, since it is usually curable, is an acceptable price to pay for protection from the more virulent forms of cancer. Others, however, insisted that skin cancer was not necessary to achieve this protection and that moderate and careful exposure could yield the same benefit. As you will see in future chapters, there is presently a renewed interest in the investigation of sunlight as a cancer preventive.

Psychiatrists have also been using light to treat depression.

The pineal substance, melatonin, implicated earlier in sexual functioning, has now been found to have an effect on our mood. Some people suffer winter depressions, and psychiatrist Alfred Lewy has found that added doses of light during this season can relieve the emotional state.[1]*

Increasingly, we are discovering that artificial alternatives to sunlight, such as the fluorescent lights in our schools and offices, may be as carcinogenic as the genuine article. A study reported in the August 1982 issue of the *Lancet* revealed that people working in offices under fluorescent lights were more likely to have melanoma than those who worked outdoors.[2] For years, the increasing incidence of this life-threatening cancer among office workers has been puzzling. Perhaps the lack of sunlight, in combination with the presence of fluorescent lighting, is responsible.

HEALTH IS MORE THAN SKIN DEEP

By tracing sunlight's fall from favor over the last several decades, it can be seen that a single organ, the skin, has received more attention than that paid to all other biological systems combined. In future chapters the influence of light on numerous other organs besides the skin will be discussed, but these effects are either only recently known or, as in the case of research from the early decades of this century, forgotten. It is with our exteriors, after all, that we greet the world. And if all is vanity, our special concern with the appearance of our skin is easily understood. On the average, we are more interested in knowing the effects the sun may have on skin cancers and wrinkling than, say, its role in modifying pineal hormones and blood leukocytes.

When we restrict our attention to a single organ and ignore the rest of the organism, we run the risk of making serious errors. The wrinkling of the skin can be slowed considerably by one's staying out of the sun, but at what cost to other or-

---

*Notes beginning on page 206 refer the reader to studies that can be found in the contemporary scientific literature.

gans? Dermatologists and molecular biologists have specified in great detail what happens when we allow our skin to burn, and much of it doesn't sound too pleasant. But the more extreme professional opinions, if taken seriously, would lead the most prudent among us to seek shelter in caves.

Harold Blum, for example, who for years has studied the effects of sun on skin, recently said that solar-induced skin cancer formation is a continuous process that begins with the first exposure to UVR. And Arthur Giese, Professor Emeritus of Biology at Stanford University, speculates on the delay that occurs between exposure to UVR and skin cancer. In laboratory work with mice, he notes, animals commonly die before UV-induced tumors appear. (This is due to the short life span of the mouse, not to the lethal effect of UV exposure.) Giese believes that it is possible, were the mouse to live long enough, that a single dose of UV radiation could produce tumors.

For the person who is quite susceptible to this type of cancer, or for those of us who expect to live awhile longer, this might be cause for alarm. Is the danger of the first exposure to UV like the problem of the predisposed alcoholic who, in the eyes of the temperance worker, proceeds inexorably to the disease of alcoholism after taking that first drink? Should we not take chances and avoid sunlight? To be scared away from liquor cannot produce many negative consequences, but to fear and avoid sunlight will.

The sun is getting a black eye. The coup de grace is delivered by those who claim that we can live without exposure to sunlight, that we can artificially simulate all its benefits. We can synthesize vitamin D artificially, but, as we shall see, its function in the body is not the same as with vitamin D synthesized naturally when sunlight falls on the skin. And artificial UV rays can cure rickets, but what of the benefits missed when other wavelengths are not present? What of sunlight's beneficial effects on liver function, heart, blood cells, and physical strength, to mention only a few recent discoveries?

Our bodies are made of the same elements as the sun (with the exception of inert helium, which is not found in animals). We are all recycled star stuff. Our evolution as humans over the past two or three million years has been guided in large measure by the sun. Living without it would lead to serious consequences.

It is clear that we cannot close the books on the beneficial aspects of sunlight but that we should also have a healthy respect for its potential harm. While dermatologists have not erred in their warnings about the damage that excessive exposure can do to the skin, it is time that we recognize that skin is but one organ of the human body and that there is such a thing as moderate exposure. Endocrinologists, hematologists, ophthalmologists, and psychiatrists, among others, are beginning to reexamine sunlight's influence. The evidence they present suggests, as we will see throughout this book, that sunlight is literally an essential nutrient, as important to good health as food, water, and air.

The sun *is* dangerous. And the sun *is* beneficent. But to come too close to either of the extremes—to bake in it endlessly or to avoid it assiduously—is to invite trouble.

2

A SUNLIGHT PRIMER

We all know *what light is; but it is not easy to* tell *what it is.*
— Samuel Johnson in 1776, according to Boswell

THE SUN, a giant, swirling ball of gases, provided both the energy and the material for life on earth. Practically all of our energy originates with the sun, from its warming rays to the fossil fuels we burn. Like the sun, we are largely hydrogen, and the other building blocks of life—carbon, oxygen, and nitrogen—are also present on the sun.

But the phrase "on the sun" is misleading. The sun has no surface, and the portion that emits the light we see and the heat we feel is largely a vacuum. The visible portion, the photosphere, is surrounded by the chromosphere, consisting of small jets of flame too dim to be noticed in daylight. The corona, the farthest extension of the sun, is visible only during solar eclipses when the moon blocks the brighter central portions. Scientists have traveled thousands of miles to be at the location of eclipses for the rare opportunity to study the corona, but today's astronomers can observe it by screening the inner layers of the sun with disks.

The Solar Reactor

As gases sink from the photosphere to the center of the sun, its core, they become denser, and the solar body takes on the characteristics of an atomic reactor. Under great pressure, hydrogen atoms are smashed together and the end result of this fusion process is the conversion of hydrogen to helium. Here is where energy is created. The newly formed helium has a

smaller mass than the original hydrogen, and though the resultant mass is only a fraction of a percent less than the original, the huge amount of energy thus released is expressed in
Einstein's famous equation: $E = mc^2$ (energy equals mass times
the square of the speed of light). This energy rises from the
sun's core and is finally released at the photosphere as light
and heat. After giving up this heat and light, the gases again
sink to the core, going back for more.

The sun is expected to serve us in much the same fashion
for another five billion years. Then it will begin to expand,
becoming what astronomers call a "red giant," and life on earth
will disappear in the baking heat. Following this, the sun will
collapse and become a "black dwarf," dead and dark.

The Solar Spectrum

Energy from the sun is emitted in the form of electromag-
netic radiation. This radiation is considered by physicists to
behave both as separate packets of energy, known as quanta,
and as waves of varying length. It serves our purposes best to
think of the sun's radiation in terms of waves whose lengths
(measured from crest to crest) vary considerably, ranging from
over a thousand meters to one-ten-trillionth of a meter. The
longest waves represent the radio waves, the shortest the
cosmic rays, gamma rays, and X-rays. It is the midrange
wavelengths, those extending from the ultraviolet (shorter) to
the infrared (longer) that concern us. Between these two ranges
lies the visible spectrum, which includes wavelengths between 400 and 720 nm. (A nonometer, abbreviated nm, is a
billionth of a meter.)

The Earth's Atmosphere

Eight minutes after escaping the sun's photosphere, radiant
energy reaches the earth. We receive about one-billionth of
the total energy released, but the energy is quite different
from that which travels through the 93-million-mile vacuum
between the earth and sun. Our atmosphere is very selective
about what it lets through.

This was not always so. Before there was life on our planet, the atmosphere was largely methane, ammonia, and water, a mixture of gases highly toxic to life as we know it. Furthermore, this primitive atmosphere allowed large amounts of UVR to pass through and reach the earth. This shortwave radiation is very injurious to the cells of all living material, and something had to change before life was possible. That something was the addition of oxygen and its by-product, ozone, to the atmosphere. But oxygen is a *product* of plant life. How was life ever established under such unfavorable circumstances?

The protection afforded by the seas allowed the development of life forms that could convert energy from the sun itself. Through photosynthesis, plants evolved in the seas at a depth at which the harmful UV was filtered out—about thirty feet. And the photosynthesis that sustained underwater plants had a by-product that was essential in the development of earth-dwelling organisms—oxygen. As our atmosphere became oxygen-rich, the stage was set for the colonization of the rocky, barren land.

While UVR can destroy life, it can also alter it. Ultraviolet radiation acts on the genetic material of organisms and creates changes, ones that are usually lethal when the organism tries to reproduce itself. Occasionally, however, these mutations are beneficial, resulting in improved survival capabilities. Ultraviolet radiation is both killer and creator. While many organisms die from exposure, sometimes a few, new, better-adapted ones evolve and thus UVR contributes substantially to the evolution of species.

THE MODERN EARTH: DAYLIGHT AND OTHER PHENOMENA

The sun-related phenomena we see on earth—daylight, blue skies, rainbows, sunsets—result from the interaction of the sun's radiation, the earth's atmosphere, and our own visual system. Two constituents of the atmosphere limit the solar radiation reaching the surface of our planet. At the short end of the spectrum, atmospheric ozone absorbs nearly all radiation with wavelengths shorter than 290 nm. Thus, the cosmic rays, gamma rays, X-rays, and most of the UVR never reaches us. At the longer end of the spectrum, atmospheric water vapor absorbs most of the radio waves and a large proportion of all wavelengths greater than 1,000 nm in length.

Our eyes are sensitive to an even narrower portion of the spectrum. Our visual system is attuned to the wavelengths between 400 and 720 nm and is most sensitive in the green-yellow range (555 nm). Since the peak of solar radiation reaching the earth lies in the vicinity of 450 to 500 nm, our eyes are fairly well attuned to the sun's energy. But is at the outer limits (wavelengths shorter than 400 nm and longer than 720 nm) that our eyes fail us, and instruments had to be developed to sense solar radiation outside of the visible spectrum.

Rainbows

Initially, human eyes were the only instruments capable of detecting solar radiation other than heat, and it was the dramatic visual phenomena that first attracted human attention. Rainbows, those ephemeral creations of reflected and refracted light, offered a clue to early scientists searching for the nature of sunlight. The colors of the visible spectrum were all there, with red on the top and blue on the bottom. Sir Isaac Newton was the first to discover, in the late seventeenth century, that he could simulate the rainbow's spectrum by passing light through a glass prism. As elusive as the rainbow's pot of gold was (since no matter where you stand you are always in the center), the real treasure to be gained was the discovery of the various wavelengths and their behavior. Red light, scientists concluded, must have longer wavelengths since it is bent less when passing through a refracting medium, whether raindrops or a glass prism. Blue light, with its shorter wavelengths, undergoes more of a deviation from its original path.

Blue Skies

Though less dramatic and more common than the rainbow, the blue color of the sky on a sunny day has long fascinated people and offered further clues for those trying to discern the nature of sunlight.

Sunshine is scattered in all directions when it encounters air molecules or small dust particles. Due to its short wavelength, blue light (as well as the invisible UVR) is scattered more than are the longer wavelengths. Because of this scattering, much light in the blue range that would normally pass above the earth's surface is reflected downward, thus giving us a blue sky. The deepest saturation of blue color is at 90 degrees from the sun, a fact we can verify by noting the intense blue dome of the sky at sunset. But to speak only of blue sky is inaccurate, since the preferential scattering of blue light occurs in the air all around us. We are, in fact, immersed in a sea of blue light. Similarly, and of great relevance to our health, UVR is also all around us.

Sunset and Sunrise

At sunset and sunrise, while the dome of the sky may remain blue, the horizon often turns shades of red and orange. At these times both the sunlight and the blue scattered light have farther to go in the atmosphere, and since the blue light is again scattered in all directions, the longer red wavelengths penetrate directly and impinge on our senses, giving the sky its brilliant sunset colors.

THE DISCOVERY OF ULTRAVIOLET RADIATION AND ITS BEHAVIOR

The existence of UVR was not known until 1801 when Johann Ritter, a German physicist, demonstrated that there was radiation beyond the end of the visible spectrum that was capable of producing a chemical reaction. Ritter knew that visible violet light had the power to reduce silver chloride to silver. When he placed some silver chloride just outside the violet range of a prism-created spectrum, the same chemical reaction occurred. Obviously there were invisible rays beyond the violet.

The UV range of solar radiation has such a profound impact on organisms that scientists have further divided it into three bands—UVA, UVB, and UVC. The longest and biologically least active type of ultraviolet, the UVA, has wavelengths between 320 and 400 nm and forms the part of the spectrum that lies closest to the visible portion. The UVA rays can produce a tan of short duration, but do not cause sunburn or a long-lasting tan except in fairly large doses.

The middle range of ultraviolet, the UVB, lies between 286 and 320 nm and can produce both sunburning and a durable tan. This is the band we will be most concerned with. Elaborate instrumentation has been developed to measure UVB, which is fortunate since we cannot see UV radiation and have no way of sensing it until our skin begins the sunburning process. Through the use of instruments, scientists have learned that UVB radiates from the sky as well as from the sun. Clouds, therefore, offer little protection from it, the vapor in the atmosphere serving merely to scatter UVB while absorbing very little.

The UVR with the shortest wavelengths, the UVC, lies between 100 and 286 nm. While the biological effect of this active radiation is considerable, we do not need to be concerned with it since the earth's atmosphere—and particularly its ozone—absorbs almost all radiation with wavelengths shorter than 286 nm. In actual field conditions in Houston, the shortest wavelength ever recorded was 294 nm.

The three bands of UV together comprise about 5 percent of the total solar radiation reaching the earth. Beyond the longer end of the visible spectrum, in the region of the infrared (between 1,000 nm and 1 *millimeter* wavelength), is found 40 percent of the solar radiation encountered on earth. But as with the UV, this radiation cannot be seen. It can, however, be felt as heat, and as with UV, Newton's prism was used to detect the existence of infrared radiation.

THE DISCOVERY OF INFRARED RADIATION AND ITS BEHAVIOR

In 1800, Sir William Hershel, a German-born English astronomer, placed a thermometer just beyond the visible red of the spectrum and registered heat in the area. This demonstrated the existence of invisible radiation beyond the longest limits of the visible spectrum.

While UVR is only *scattered* by water vapor in the atmosphere, water vapor *absorbs* infrared radiation, thus accounting for the cooling effect of clouds positioned between us and the sun. And shade of any kind blocks the infrared.

ATMOSPHERIC OZONE

Solar radiation is not compatible with life as we know it. The sun's cosmic rays, X-rays, gamma rays, and the shorter UV rays kill. Our salvation is the earth's atmosphere, a filter that prevents most of these harmful radiations from reaching us.

For a long time, scientists were puzzled at the sharp limit of the solar spectrum on earth. The spectrum was continuous down to the UVB range and then suddenly ended—shorter wavelengths could not be detected even with instruments. Sir Walter Hartley correctly concluded in 1881 that ozone was responsible for the truncation of the spectrum. Ozone absorbs practically all of the radiation shorter than 286 nm and, as was later discovered, much of the radiation with wavelengths shorter than 330 nm as well.

Ozone, or O_3, is concentrated in a layer between six and thirty miles above sea level and has a maximum concentration at an altitude of fifteen miles. Even at this level it constitutes but one in four million parts of the atmosphere. If all the ozone in this layer were compressed, it would form a ribbon around the earth only one-tenth of an inch thick. Ozone is indeed a rare and precious molecule.

An interesting chain of chemical events results in the formation of ozone. In the upper atmosphere, energetic light rays with a maximum wavelength of 243 nm strike *molecules* of oxygen (O_2) and cause them to split into two oxygen *atoms* (O). The atoms of oxygen then combine with O_2 and ozone is formed. Ozone itself is decomposed when it does its job of protecting us by absorbing the short rays, and the cycle begins anew. We find ozone concentrated between six and thirty miles above sea level because at higher altitudes there is too little oxygen to be dissociated into oxygen atoms and get the ball rolling. Below this level, existing ozone has done its work and too few short rays capable of dissociating ozone get through. So the scarcity of oxygen at high altitudes and the scarcity of short-wavelength radiations at low altitudes limit the production of ozone. As we will soon see, however, ground-level ozone can occur as sunlight strikes the exhaust products of internal-combustion engines.

The amount of ozone in the atmosphere depends on several factors. There is a seasonal variation, with the maximum during April and the minimum in October. And because of the nature of our planet's air currents, ozone from the tropics is carried to the higher latitudes, making it more plentiful in nontropical areas. Finally, the concentration of ozone near the ground is significant in cities, due primarily to automobile-exhaust gases.

Along with the awareness of ozone's importance to life came a concern over the possible destruction of it through human activity. Nitrogen oxides (NO)—produced by such things as automobile exhaust, commercial fertilizers, stratospheric aircraft, and atmospheric explosions of nuclear weapons—react with ozone to produce nitrogen dioxide (NO_2) and oxygen. Ozone is thus destroyed.

$$NO + O_3 = NO_2 + O_2$$

Unfortunately, there is more to the story. Nitrogen dioxide further reacts with atomic oxygen, and more nitric oxide is regenerated, ready to again react with and break down more ozone.

$$NO_2 + O = NO + O_2$$

So nitrogen oxides can start a chain reaction of increasing ozone destruction.

The stratosphere, beginning seven miles up and constituting the outer portion of the atmosphere, is the home of ozone. The interaction of short-wavelength rays with oxygen, resulting in our protective ozone, could go on indefinitely. In the natural course of events, the nitric oxide that destroys ozone does not occur in the stratosphere. But when, through human activity, nitric oxide is introduced into the stratosphere, it remains to perpetuate its cycle of ozone destruction for between one and three years. The stratosphere is a very unforgiving part of our atmosphere, and substances introduced into it remain, since there is little mixing with neighboring layers.

What are the effects of a reduction in atmospheric ozone? It is estimated that a 1 percent reduction in the ozone level would increase the amount of UVB reaching the earth by 2 percent, and that for each 1 percent decrease in ozone, the various skin cancers would show a 5 to 10 percent increase in incidence.

One of the reasons American citizens fought the development of fleets of SSTs was their potential for destruction of the ozone layer. Similarly, the use of chlorofluoromethanes (trade name, "Freon") as propellants in spray cans was severely curtailed in 1977 when it was learned that they rise unchanged to the stratosphere, where UVC rays convert them to chlorine atoms and other compounds. The chlorine atoms break down ozone in a manner similar to nitric oxide. Though the use of Freon in spray cans was banned, it is still widely used as a refrigerant and foaming agent.

In 1975 the National Academy of Sciences released a report entitled "Long-term Worldwide Effects of Multiple Nuclear Weapons Detonations."[1] Nuclear explosions generate such high temperatures that oxygen and nitrogen in the air combine to form our old enemy, the nitrogen oxides. The resulting ozone destruction, says the report, would have devastating consequences. If half the nuclear arsenal of the United States and the Soviet Union were detonated in a nuclear exchange, the resultant 50 to 75 percent reduction in ozone would have a more disastrous effect than the immediate radiation of the blasts. The damage caused by the explosions would mainly effect the area of the strikes. The effects of atmospheric ozone destruction would eventually destroy the world.

So far, in discussing the ramifications of human activities

for the ozone layer, we have dealt only with speculation. What happens when we actually measure the atmospheric ozone concentration over a period of years? Surprisingly, between 1963 and 1970 the measured level of atmospheric ozone *increased* by 5 percent. In the face of growing numbers of automobiles, increasing industrialization around the world, and the cumulative effect of Freon from spray cans, how is this possible? Though no one knows the answer for certain, some scientists believe that the ozone increase was due to a recovery from the effects of earlier nuclear testing. Between 1952 and 1962, a large number of atmospheric nuclear tests were conducted by the Soviet Union and the United States, and the large concentrations of nitric oxides produced in the stratosphere may have decreased ozone levels. The increase reported between 1963 and 1970 probably represents a recovery from these earlier depressed levels.

In 1982 the National Academy of Sciences reported that, to date, no further increase or decrease in ozone attributable to human activity has been found.[2] In what might be included under the heading of "good news–bad news" the Academy reported that while ozone in the stratosphere has decreased by one-half of one percent in the past decade, the amount at ground level has increased by the same amount. The latter is due to an increase in pollutants, mainly automobile exhaust, and their conversion to smog. If these pollutants continue to increase, the previously predicted decrease in stratospheric ozone will be largely offset, so that by the end of the next century, the net decrease will probably amount to no more than 1 percent. Actual measurement of UVR at ground level reveals no change over the last decade. But though UV intensities have not changed, the redistribution of ozone may cause temperature changes, ones that could have a significant effect on the world's weather.

PHOTOCHEMICAL SMOG

Atmospheric ozone is essential to life, but we could live without the ozone we help produce at ground level. Ozone, along with carbon monoxide (CO), is one of the components of eye-burning, lung-irritating smog. When the atmosphere contains concentrations of nitrogen dioxide, formed primarily in internal combustion engines, the conditions are right for smog formation. When UVR strikes this nitrogen dioxide, it is converted to ozone, but unlike ozone in the upper atmosphere, at ground levels it is harmful to health. Together with other compounds in the atmosphere, it forms what is called photochemical smog. As the term implies, sunlight is a major factor in the creation of this smog, and it is thus most likely to occur in sunny places such as Denver, Los Angeles, and Phoenix.

Depending on its constituents, smog varies in color from blue (when particulate matter is very small), to yellow, brown, or black, depending on which wavelengths the suspended particles reflect. Compare this to the appearance of fog, which is white in color because the water droplets forming it reflect all colors. (White light consists of all the colors of the visible spectrum.)

Photochemical smog is a severe health threat which can significantly shorten the lives of those already suffering from respiratory ailments. It is at its deadliest in combination with a climatic event known as a temperature inversion. Inversions occur most frequently in desert areas where ground heat radiates readily to the clear, dry, nighttime sky. In desert basins and valleys, cold air slides down the mountainsides at night and pools in the lower elevations. The warm air, heated by the previous day's sunshine, rises, and an unusual phenomenon occurs—the warm air rests on top of the cooler air. This effectively forms a cap over the valley air, trapping pollutants and photochemical smog for days at a time until a wind moves it away or rain rinses it out.

But such occurrences are not unique to desert regions. The worst smog-related catastrophe occurred in London, of all places, during December 1952. In this case, a dense fog settled and created a temperature inversion. Smoke, dust, and fumes were trapped for days and visibility was only a few inches. Four thousand people died and uncounted others suffered permanent injuries as a result.

NATURAL AIR POLLUTION

Sun-blocking air pollution can also be the result of natural phenomena such as volcanic eruptions and forest fires. Perhaps humans take too much of the blame for the particulate matter, also known as aerosols, in the atmosphere. Volcanic dust is by far the most plentiful of them all. In fact, the quantity of all naturally produced atmospheric pollutants is eight times the amount produced by humans.

The recent eruptions of Mount St. Helens in the State of Washington and El Chichón in Chiapas, Mexico, have reminded us of something that Benjamin Franklin noticed in 1783—that volcanic eruptions can reduce the amount of solar radiation reaching the earth. Franklin observed, after a volcanic eruption, that light passing through a magnifying glass would no longer set a piece of paper on fire.

Though El Chichón put only one-sixth the material in the air as did Mount St. Helens, its effect on world weather is likely to be much greater. El Chichón blew straight up, delivering its ash directly to the stratosphere, whereas Mount St. Helens projected its ash at a lower angle. The enduring atmospheric cloud produced by El Chichón is one hundred times denser than are the remains of Mount St. Helens and may lower global temperatures by as much as one-half degree Fahrenheit during the next few years.

THE DIRECT EFFECT OF SUNLIGHT ON ORGANISMS

To have an impact on the cell of an organism, sunlight must first be absorbed instead of being reflected or transmitted (passing through without change). Different molecules absorb only certain wavelengths of radiation, and as they do they are activated or excited. A chemical change in the molecule may be the result, or the radiation may be reemitted as radiation of a longer wavelength or as heat.

Vision is a prime example of a chemical change induced by light. As light strikes our retina, pigment in the receptor cells of the retina absorbs it and communicates the information to the brain by means of electrical and chemical changes.

By knowing which wavelengths a compound or molecule will absorb, scientists have been able to determine the specific mechanisms of some of sunlight's effects. For example, it has long been known that UVB radiation kills bacteria, and it has been used therapeutically for this purpose. But just how is this accomplished? Since bacteria's genetic material, the DNA, absorbs UVB radiation and its protein does not, it can be concluded that bacteria cell destruction is a result of DNA damage and not the effect of UVB radiation on protein.

The Skin

After the visual system, the skin exhibits the most obvious change when exposed to sunlight. Laypersons and scientists alike have focused most of their attention on sun-induced changes undergone in the skin. Let us look, then, at some basic information on this organ.

The outer layer of the skin is called the epidermis, and the layer below it the dermis. The epidermis is further divided into the stratum corneum (the portion we can see) and just beneath it, the basal cells. Between the dermis and the epidermis lie a network of blood vessels.

Whether sunlight acts upon a certain layer of the skin depends on the wavelengths present. The shorter UVB radiations don't penetrate far at all, about 90 percent of them being absorbed by the epidermis. The longer UVA rays have a greater penetrating power, and a substantial proportion get through the epidermis and act on the blood vessels and dermal layer. As we enter the visible portion of the spectrum and the wavelengths get longer and longer, penetration is deeper. The longest rays of the visible spectrum can pass not only through both layers of skin, but through the whole body as well. Holding a bright flashlight against your hand in the dark will demonstrate this point. The light from a flashlight is mostly in the red region of the spectrum, and we can see it shining right through, showing even the shadows of our bones.

Sunburning and Tanning

Surprisingly, it is not the radiation with the greatest penetrating power that affects the skin most significantly. Remember that radiation must be absorbed before it can cause any chemical change within the cell. The red light passes through with little effect. Infrared radiation from the sun, with its relatively longer wavelengths, can cause an immediate flushing due to heating and a dilating of blood vessels, but this is a fleeting phenomenon and passes soon after we get out of the sun. Real sunburn, or erythema, is produced primarily by the action of UVB rays on the epidermis, sometimes in conjunction with UVA radiation acting on the dermis. It works like this: when the epidermis absorbs enough UVB radiation to cause a barely noticeable reddening, skin cells are damaged. (The degree of exposure necessary to cause this redness is called the minimum erythemal dose, or MED.) These damaged cells release a substance which causes the blood vessels to relax and increase their diameter. As a result of the augmented blood supply, the skin turns red and also feels warm. With further MEDs of sunlight, the sensory receptors in the skin are injured, and touching the affected area causes pain.

The problem is that even though our skin may flush upon first exposure to the sun, actual sunburn does not appear until from four to fourteen hours later. It may not reach its fullest intensity for twenty-four hours, and MED is usually determined after this period of time. If we have received one or only a very few MEDs, the changes begin to subside shortly after this peak is reached, and the burn is gone in about four days.

Since it takes so long to feel the full effect of sunburn, it is common for a person's skin to absorb many MEDs of UVR without feeling any pain or noticing a very significant redness right away. If between five and ten MEDs are received, the red coloration later becomes pronounced, the pain intense, and blistering can occur. This too subsides and the skin peels in a few days.

As noted, UVB is the radiation most responsible for sunburning, and research has shown that the most active wavelength for producing these effects in the skin is at 297 nm. Longer UVA radiation, while also capable of causing a burn, is much weaker than UVB, requiring about one thousand times the dose to produce the same erythema as UVB. So UVA is not too effective in causing a sunburn, but there is evidence that it can add slightly to the effect of UVB, and thus a worse burn may be the result of exposure to both bands.

Most of us, however, do not seek a sunburn—what we are after is a tan. How does tanning differ from the sunburning process?

Actually, burning and tanning share much in common. As with sunburning, there is often an immediate tanning which appears shortly after exposure to sunlight as well as a delayed, longer-lasting tan. Immediate tanning occurs within minutes of exposure to UVA or even visible light, and involves a darkening of the pigment already in the skin. It can happen when the sun shines through window glass. One is unlikely, however, to get a real sunburn or a long-lasting tan through a window, since the glass transmits only radiation longer than 320 nm, the upper limit of UVB.

The immediate tan, like the short-term sunburn, is usually gone after a few hours (though it may last for as long as thirty-six hours). A delayed tan is the result of cell injury by UVA or UVB and can occur without erythema.

About ten hours after exposure, the delayed tanning process begins though the tan does not reach its peak for from four to ten days. As you can see, what most people call a tan after a day in the sun is actually either an immediate tan or a sunburn. The long-lasting tan involves photosynthesis, the development of new melanin pigment which is taken up into the skin cells, causing the skin to look dark. At the same time, the epidermis thickens. The development of new pigment and this thickening of the skin are the factors that differentiate immediate tanning from delayed tanning. But even the long-lasting tan must fade as the pigmented cells gradually flake off, restoring the skin to its former color after a few months.

INDIVIDUAL DIFFERENCES: SKIN TYPES

People differ in the amount of time it takes to sunburn or tan, and the type and color of skin they were born with is the most important factor. Skin varies in color from the deepest black of equatorial people to the virtual white color of the albino. (Normal Caucasians are not white, but pink, gray, and shades of brown, etc.)

Our skin color is actually a mixture of red, blue, yellow, and brown. In the dermis, oxygen-carrying blood in the capillaries creates a red color while deoxygenated blood in the veins appears blue. In the epidermis, carotenoids (substances found in carrots as well as the skin) create a yellow color. Melanin, also in the epidermis, is a brown-colored pigment and determines one's skin color more than all the other factors.

As we know, skin color changes upon exposure to sunlight, so we have to distinguish constitutive skin color (color in the absence of any solar radiation) from inducible skin color (the result of tanning). What is not commonly known is that factors other than heredity can enter into constitutive skin color. For example, exposure to the sun over a period of many years can darken one's constitutive skin color. The skin becomes darker and remains thus without additional sunning. We can observe this among retired farmers, seamen, or cowboys who, while no longer outdoors a great deal, retain a skin color that is darker than when they were young.

Your Skin Type

The Food and Drug Administration has developed a list of six skin types. You can determine your skin type from the following descriptions:

SKIN TYPE I. Always burns easily; never tans. Persons in this category are very sensitive to sun. They often have red hair or freckles. Interestingly, while those with other skin types have a kind of melanin pigment called eumelanin in the skin and hair, redheads have pheomelanin. Pheomelanin is highly unstable in sunlight and when exposed to UVR undergoes changes which may be damaging to skin cells. Research on pheomelanin is under way, and at present we can only say that its role in skin cancer formation presents an interesting hypothesis.

Not all type I's are redheads, however. People of Celtic ancestry (from England, Ireland, Scotland, or Wales) also fall into the category. Their burning and tanning history is similar to redheads, and the amount of melanin in the skin is small. The melanin that is present is not sufficiently uniform to allow an even tanning, so type I's can never attain the tan that would give them some protection on future days in the sun. Because of the lack of pigmentation, vitamin D is synthesized most readily in the skin of this group.

SKIN TYPE II. Always burns easily; tans minimally. Persons in this category are also sensitive to the sun. Skin is fair and their hair is light colored. Eyes are generally blue or some other light shade such as green, and ancestors are typically from northern latitudes, usually northern Europe. (Light hair and eyes are found only among people of European descent.) Type II's must be careful in the sun, but since they can tan minimally, the slow acquisition of a suntan could reduce the likelihood of a burn upon future exposures.

SKIN TYPE III. Burns moderately; tans gradually and uniformly to a light brown color. Persons in this category have a normal reaction to the sun. They are usually of European ancestry and, while Caucasians, are somewhat darker than type II's. As we enter into the normal range of skin types, the incidence of skin cancer drops markedly. The majority of skin cancers are found in persons with type I or type II skin.

SKIN TYPE IV. Burns minimally; always tans well to a moderate brown color. As with the previous type, persons in this category are also considered to have a normal reaction to the sun. They are frequently Mediterranean-type Caucasians of rather swarthy complexion.

SKIN TYPE V. Rarely burns; tans profusely to a dark brown color. The FDA labels persons in this group as minimally sensitive to sun, though it should be noted that extended expo-

sure can still cause problems. Ancestry includes Middle Eastern regions and Latin America.

SKIN TYPE VI. Never burns; deeply pigmented. Persons in
this category are also considered relatively insensitive to the
sun. Most blacks are in this category.

The FDA's use of the terms "relatively insensitive" and
"never burns" should be used with some caution. While the
profuse and well-distributed melanin pigment in black skin is
one of the best sunscreens known, blacks may also suffer from
overexposure to the sun. While sun-induced wrinkling is far
less common in them, type VI's whose occupations keep them
outdoors a good deal of the time can show the same furrowed
cheeks and foreheads seen in whites who have had too much
sun. As noted previously, blacks are more likely to suffer the
consequences of too little sun than are whites, and the black
person working indoors in a northern city is much more likely
to develop a vitamin D deficiency and exhibit other symptoms
of sunlight starvation.

Nature's Sunscreen

Derek Cripps of the University of Wisconsin's Department
of Dermatology has determined the degree of protection provided by each of the skin types.[3] He ascertained the MED for
each type and gave each a sun-protection-factor (SPF) rating.
(This SPF rating is used the same way as the SPF number on
sunscreen products.) Setting the SPF for skin type I at 1.0,
he found the following SPFs for the six skin types:

Skin Type I:	1.0
Skin Type II:	1.7
Skin Type III:	2.5
Skin Type IV:	4.0
Skin Type V:	4.0
Skin Type VI:	9.7

Thus, relative to skin type I, a type III can receive two and
a half times the UVR before exhibiting a barely perceptible

burn. A type VI can take almost ten times the amount of UVR as a type I.

RESEARCH ON SUNLIGHT AND HEALTH

Scientists who investigate sunlight in relation to health have been more concerned with short-wavelength radiation and less interested in the longer wavelengths of the solar spectrum. The sun's infrared radiation affects our bodies through heat production, but comparatively little research has been done in this area. The visible spectrum, red through blue and violet, stimulates visual processes and other body processes controlled by light coming through the eyes. But again compared to research in the UV range of the spectrum, relatively little has been done to study the effects of these wavelengths of sunlight on human health.

In recent decades, there has been a trend in the research on sunlight and health. Early studies often employed artificial light sources to simulate the effects of sunlight, and many researchers used UVC irradiation and noted its impact on organisms. As noted in Chapter One, the effects were dramatic, leading sometimes to cures and other times to pathological developments. But any attempts to extend these findings to questions about sunlight were ill-conceived since UVC radiation does not reach the surface of our planet in any appreciable amount.

Interest then grew in UVB, since this had more relevance to human health here on earth. The wavelengths responsible for sunburn, tanning, and skin cancer were identified (the most effective being, in each case, around 297 nm). The spectrum of concern then widened to include the longer-wavelength UV rays, the UVA.

At first, UVA radiation was termed "beneficial ultraviolet" since it could induce pigmentation and thus protect against future exposure to UVB. The coloration induced by UVA is primarily immediate pigmentation and not the type that leads to a lasting tan. But if, on a given day in the sun, one darkens immediately, this might afford some protection from UVB for the rest of the day. The complete story on UVA was neither so simple nor so positive. One researcher threw in a scare which proved to be a false one. His studies showed that UVA, far from being protective, actually made the skin more vulnerable to UVB and was thus an important factor in the formation of skin cancer. But the research was flawed in a manner that characterizes too many studies of sunlight and health: the UVA and UVB given to subjects were way out of proportion to that present in the sun—the UVA was much more intense than that found in sunlight.

We no longer refer to UVA as "beneficial ultraviolet," but neither is there any reason to believe that it has special cancer-causing properties. Current research on UVA suggests that it is relevant to health in three ways. First, it can cause the immediate pigment darkening referred to earlier. Second, because it penetrates deeper than UVB, going through the epidermis and into the dermis of the skin, it may well be implicated in wrinkling. The full picture of solar-induced wrinkling includes both excessive UVA, which causes elastosis (a breakdown of elastic tissue) in the deep dermis, and UVB, which adds to the effect by damaging the epidermis. Wrinkling will be discussed more fully in Chapter Four.

The third way that UVA affects health is by *contributing* to real sunburn and long-lasting tanning. The shorter UVB rays are far more effective in this regard and it takes a thousand times as much UVA to produce the same result. But since the ozone layer filters little of the UVA, it is much more plentiful in the sunlight reaching earth than is UVB. If weather conditions are right, UVA can be about 10 percent as effective as UVB in producing burning and tanning.

John Parrish and his associates at the Harvard Medical School have shown that UVA is simply a weak relative of UVB when it comes to sunburning and tanning.[4] They produced a barely perceptible redness in skin with several combinations of UVA and UVB. In each case, the UVA and UVB had an additive effect. For example, when ¼ MED of UVA was given, it took ¾ MED of UVB to produce redness. And when ¾ MED of UVA was given, ¼ MED of UVB was required before erythema appeared. (Note we are dealing with MEDs here—remember that it takes a thousand times the amount of UVA compared with UVB to produce redness.)

Among dermatologists and others concerned with the effects of sunlight on skin, both types of UVR that reach the earth are currently of interest. Sunscreen manufacturers have kept pace with these recent developments and now market products that filter out both UVA and UVB.

The spectrum of interest for most of the conventional scientific community stops here. While there is some interest in visible light and infrared radiation, the controlled research on the effects of these wavelengths on human health is a drop in the bucket compared with the research on UVR. By necessity, then, a substantial proportion of this book is concerned with assessing and controlling our exposure to UVR.

But before looking at our problems with the sun, let's examine the benefits it confers on human health. Some of the research in the next chapter is quite recent, and I, for one, hope that this is the beginning of extensive programs investigating what most of us intuitively feel—that the sun is our friend.

3

THE SUN AS FRIEND

If only the sun would come out, I would have the score finished in no time.

—Richard Wagner

WHY IS our mood so elevated when the sun shines, so depressed when it's cloudy? There may be some purely aesthetic or psychological reasons. A world of shadow and light is more beautiful than a drab, overcast world. In the sun we may feel that we are at work on a tan we can prize, or we may associate sunshine with pleasant outings away from work and worry, rain and clouds with the postponement of these respites. But sunlight also affects our mood in a more direct manner, by producing changes in the body that are healthy and exhilarating. And evidence is accumulating that strongly suggests that light is an important environmental factor, along with food, water, and air, in regulating our bodies' physiological functions.

But discovering the specific ways in which sunshine helps us is not as easy as it sounds. For most people, few days go by without at least some exposure to the sun. And though a day may be cloudy, a portion of the sunlight gets through to us nonetheless.

Some people, however—Arctic inhabitants and underground miners, for example—go for varying periods of time during which they are totally deprived of sunlight. And the blind, while benefiting from light that reaches their skin, often lack that portion of the optic system that is stimulated by light and sends messages throughout the body. Do these varied groups experience any problems associated with their deprivation?

ARCTIC INHABITANTS

To discover the effects of lack of sunlight, the Arctic environment furnishes one of the best natural laboratories we could hope for. Beginning in the early autumn, days in these latitudes become rapidly shorter and the zenith of the sun sinks lower in the southern sky. By December, there are only a few hours of twilight during the "day." January brings total darkness with nothing but a brief red glow on the horizon at noon, and UVR is nearly absent at this time. Even during autumn and spring, when the sun does rise above the horizon, the Arctic dweller's heavy clothing prevents the weak UV rays from reaching the skin though not, of course, the eyes. What are the effects on health?

Among scientists who work in the Arctic, a number of problems have been noted. During their long winter night, controlled studies have demonstrated such things as a decrease in sexual drive and potency, insomnia, weakness, loss of hair, and general depression and irritability. All of the above symptoms disappear upon exposure to sunlight.

We might dismiss the complaints of those who spend a few months in the Arctic laboratories and then return to their more temperate homes as problems of adjustment. They are in an unfamiliar environment, one which is hostile and bleak. Perhaps they suffer stimulus deprivation.

Those born to an Arctic existence, on the other hand, have had years to adjust to the conditions of their homeland. To them, the Arctic is neither foreign nor hostile, but simply normal. And the landscape we would call bleak is, for those who live there, constantly changing and stimulating. How does the lack of sunlight affect those who live their lives north of the Arctic Circle?

The Eskimos

In 1976 the International Symposium on Circumpolar Health was held in Yellowknife, Northwest Territory.[1] The catalog of Eskimo diseases surveyed was extensive. Very high rates of tuberculosis, rheumatic fever, hepatitis, diabetes, leukemia, cardiovascular disease, and cavities of the teeth were reported. Suicide, alcoholism, and other emotional problems were also widespread. It would be foolish to attribute all of these problems to a lack of sunlight, since several are shared by Indians of the American Southwest, a group that certainly does not lack sunshine. The Pima Indians of Arizona, for example, have a higher incidence of diabetes than any group in the world. But other groups that do suffer sunshine deprivation have several of the same disorders.

UNDERGROUND MINERS

Studies of miners deprived of sunlight are also of interest. Especially in the winter months, a worker may enter the mines before sunrise and not leave until after dark. In Chapter One the general health improvement of people who moved west in the late nineteenth century was noted. Miners, however, were a consistent exception to this rule. Many reports testify that within a few months in the mines, formerly healthy men were pale, shrunken, and broken down. Mortality from disease was very high among this group.

In England, the Royal Commission offically looked into the matter of miners' poor health in 1840. They concluded that "this Employment, as at present carried out in all districts, deteriorates the physical constitution. The limbs become crippled and the body distorted, and in general the muscular power gives way and the workpeople are incapable of following their occupation at an earlier period of life than is common in other branches of industry.

"That by the same causes, the seeds of painful and mortal diseases are often sown in childhood and youth; these slowly but steadily developing themselves, assume a formidable character between the ages of thirty and forty; and each generation of this class of the population is commonly extinct soon after fifty."

Among the common diseases reported by the commission were dermatitis, rheumatism, and miner's nystagmus. Those afflicted with nystagmus exhibited symptoms of anxiety, tremor, listlessness, and loss of equilibrium as well as the rapid, involuntary eye movements typical of the disease. This particular problem, caused by poor lighting, was easily prevented and has nearly disappeared with the advent of artificial illumination in mines.

Dermatitis, however, is still common among miners and may well be due to the absence of bacteria-killing UVR. The prevalence of colds and other respiratory ailments among underground miners has also been a continuing problem.

Meade Arbele, a writer who worked for a year in a Pennsylvania coal mine, repeatedly notes the poor physical condition of the men and women he worked with. Most of those who had been there for any length of time were "a mess inside, with ulcers, hemorrhoids, and nervous disorders." Arthritis, aching joints, and bursitis were also epidemic among his co-workers. Emotional problems, Arbele observes, were unusually frequent, and the miners' complaints of poor sex lives were constant.

As a preventive for some of these problems, miners, especially in the Soviet Union, have been given phototherapy. Over fifty years ago, however, the first such use of UV therapy with miners was introduced by a mine in the United States. In 1930, at the Bunker Hill & Sullivan Mining Company at Kellogg, Idaho, underground miners were offered voluntary exposure to "artificial sunshine" delivered by UV quartz lamps. Miners, wearing only loincloths, stood on a narrow, mobile platform which moved them through a cabinet equipped with banks of lamps. The trip through the cabinet lasted one minute and was repeated three times weekly. Though it was anticipated that colds and other respiratory ailments would be reduced by one-half, the results of this phototherapy were never reported and the experiment has not been repeated in this country. As you may recall, phototherapy, along with heliotherapy, began a precipitous decline in the 1930s as the dangers of overexposure to UVR came to light. Such mass exposure was subsequently deemed unwise.

But far more serious problems than colds and dermatitis are common among miners, and in some countries phototherapy is still used to combat them. Miners suffer extraordinarily high rates of lung diseases—silicosis from breathing silica-laden dust and black lung disease caused by breathing coal dust. These afflictions seemingly have little to do with a lack of sunlight, but Soviet researchers have reported success in the treatment of black lung disease by giving miners UV therapy. At present this relationship between lung disease and UVR is only empirical, and no one knows precisely how or why it works. Perhaps, as with early heliotherapy of the many tuberculoses, UV rays strengthen the substrate of the whole body, allowing it to fight off disease at numerous specific sites, including the lungs. Soviet researchers may give us the answer soon.

THE BLIND

Sunlight not only affects our massive exterior organ, the skin, but also enters the body via the eyes. In normally sighted people, light stimulates the optic nerve and a message is sent to one of the endocrine glands, the pineal gland. In people with cataracts or with a damaged optic nerve, however, no message is sent to the pineal. In the following pages, several studies that have uncovered differences between the physiological functioning of the blind and sighted are discussed. And equally as interesting are several changes that occur when blind persons regain their sight after successful cataract surgery.

BUT WHAT CAN WE PROVE?

Other ways of studying the effects of sunlight include comparing body functions during the day and night, noting changes from summer to winter, and discovering the health-related correlates of latitude. These techniques—as well as the previously mentioned observations of Arctic inhabitants, underground miners, and the blind—all have limitations. Factors other than sunshine vary from summer to winter—temperature, for instance. And the blind differ from the sighted in ways other than visual acuity. In most cases, the blind are less active than the sighted, and the level of mobility may account for some of the observed differences between these populations.

Similarly, miners are exposed to health hazards other than the lack of sunlight. The dust they breathe, whether coal dust in a coal mine or quartz dust in a gold mine, can have disastrous consequences for the lungs. Finally, many inhabitants of the Arctic—another unhealthy group—are financially impoverished in addition to being deprived of sunshine. The health-related correlates of poverty do not require enumeration.

If all our methods for studying the effects of sunlight deprivation on the human body have such serious limitations, what can we conclude with any certainty from this type of research? While one study by itself may not be much help in understanding sunlight's relation to health, several studies taken together allow us to make statements with more confidence. Each study with a particular group may have limitations, but all studies do not share the same limitations.

We may not know whether lack of sunlight or inhaled dust (or both) caused the generally deteriorated condition of last century's miners, and we do not know whether lack of mobility in the blind causes some of the differences between sighted and unsighted populations. But if the same physiological problem occurs in the blind *and* in miners, we have more confidence that its cause can be attributed to a sunlight deficiency. Certainly, miners do not lack mobility, and the blind do not characteristically breathe large amounts of quartz dust. These factors cannot account for observed differences in both groups. If we find a number of different populations that all have the same reactions to lack of sunlight but whose other circumstances differ widely, we may be more certain that their symptoms can be traced to sunlight starvation. Throughout this chapter, reference will be made to the problems shared by various groups deprived of sunlight.

Controlled experimental studies with animals offer the final confirmation. With animals, conditions can be varied systematically and the consequences of a single factor, sunlight, can be observed. Animals and humans differ, of course, but we also share much in common. For this reason, lower animals are often used as research subjects in this type of experiment.

SUNLIGHT AND ANIMALS

Indeed, observations of animals gave us the first clues as to the changes that sunlight could cause in organisms. Toward the end of the last century, farmers had already noticed that lack of sunlight could stimulate the production of more fat on their animals. Illumination levels were consequently lowered so these animals might gain weight more quickly. With chickens, more light increases egg production. Poultry farmers have long known that hens produce more eggs when kept under artificial lighting conditions which approximate the length of a spring day. Between fifty and eighty more eggs per laying period are so produced.

A phenomenon that has always fascinated humans is the changeable colors that various animals such as chameleons exhibit. In 1916 the American zoologist Samuel Mast showed that a plaice (a fish) changes its color when light enters the eyes. In an interesting experiment, Mast alternately put a plaice's head against light and dark backgrounds. Body color was shown to depend solely on the quality of light entering the eyes. Whether the rest of the body lay on a light or dark background had no effect on color.

But in humans, it was the discovery of the "sunshine vitamin" and its impressive implications for health that first captured attention.

HUMAN HEALTH: THE SKIN, UVR, AND VITAMIN D

As previously mentioned, sunlight has at least one positive effect on human skin—vitamin D is produced in skin exposed to UVR. Without sunlight and vitamin D (actually not a vitamin at all but a vital hormone) the body cannot efficiently absorb calcium.

Research conducted among the elderly at the Chelsea Soldiers' Home near Boston revealed that even with an adequate supply of calcium in the diet, the lack of exposure to UV rays causes a calcium deficiency.[2] In this study, all men remained indoors for seven weeks. Their ability to absorb calcium fell by 60 percent. Half of this group was then continued on a sunless regimen and the other half was exposed to artificial lights designed to simulate sunshine. Eight hours per day of exposure to these lights was the equivalent of a fifteen-minute walk in the midday summer's sun. Among this group absorption of calcium increased by 15 percent above the previously depressed level, while the sunless group's absorption fell an additional 25 percent. Other studies have shown that under-mineralization of the bones due to calcium deficiency is more common in autopsy samples collected in winter as compared to summer when UV levels are higher.[3]

The quantity of vitamin D in the blood varies directly with the total yearly UV level at different latitudes. The resident of Palm Beach has twice the amount of vitamin D as does the person who lives in Seattle, with the person in Denver falling midway between. Moreover, outdoor workers around the country have vitamin D levels which, on the average, are 50 percent greater than for indoor workers. Indoor workers' August levels are about the same as outdoor workers' February levels.

Vitamin D probably played an important role in the reports of sunshine cures coming from the early decades of this century. Much of the literature of the day listed asthenia—a general weakness and loss of strength—as improving with exposure to sunlight. Vitamin D deficiency may well have been the common denominator of those problems involving lack of strength and vitality.

Among infants and children, the need for calcium is great, and the lack of a sufficient amount of vitamin D leads to rickets, which causes the long bones to bend and the formation of nodular enlargements on the sides and ends of bones. In addition, the soft spot on an infant's head closes slowly, muscles ache, and liver and spleen can degenerate.

Sunlight, vitamin D, and calcium are also related to the quality of children's teeth. In 1938 the results of a U.S. Public Health Service investigation of the dental health of a million and a half children were presented before a medical convention. Bion East, D.D.S., reported a strong relationship between the mean annual hours of sunshine in a particular locality and the incidence of cavities—the more sunlight, the fewer cavities. Areas having more than 3,000 hours of sunshine per year reported an average of 290 cavities per 100 children. Areas with 2,600 to 2,999 hours of sunlight reported 11 percent more cavities, and those with only 2,200 to 2,599 hours of sunlight, 30 percent more. Those parts of the country having the least sunshine, fewer than 2,200 hours per year, reported 68 percent more cavities than the sunniest areas. Around the country, East found, cavities develop more frequently in winter than in summer. All in all, this is strong evidence that hard, healthy teeth are related to sunlight, probably mediated by vitamin D formation and the subsequent metabolism of calcium.

Artificial Vitamin D Versus Natural Vitamin D

The development of synthetic vitamin D (or D_2 to distinguish it from natural D_3) ended the massive threat that rickets once posed to health. Vitamin D_2 can be formed by irradiating a natural plant substance, ergosterol, with UV rays.

In the 1920s, researchers discovered that when animals are irradiated with UVR, vitamin D develops in the skin, liver, and muscle, and that these parts, when fed to children, are antirachitic (that is, they cure rickets). Moreover, eggs and milk similarly irradiated take on the same properties. Since this discovery, milk—which is poor in vitamin D—has been routinely irradiated with UVR, converting its ergosterol to vitamin D. For a while, in fact, mothers were given UV irradiation to increase the vitamin D content of the milk they fed their infants.

There is strong evidence, though, that D_2 is not as biologically effective as D_3. While the artificial vitamin can cure rickets, its effect on the total organism is much weaker than D_3. When blood samples are collected and analyzed, scientists have found that between 70 and 90 percent of the vitamin D activity in the body is due to D_3, and that this sunlight-produced vitamin is vastly more important than that obtained in the diet.[4] Thus, while vitamin D_2 can prevent the disease of rickets, it cannot take the place of vitamin D_3 in the maintenance of total health.

Sunlight and the Elderly

While the need for sunlight vitamin D is great among children, the elderly—especially those who are housebound—should also be aware of their need for sunlight. As is well-known, old people are vulnerable to what is called nutritional osteomalacia, or a softening of the bones due to vitamin D deficiency. This malady can transform a minor fall into a serious injury from bone breakage.

Research conducted by H. M. Hodkinson and his associates has revealed that, despite the name of the disease, diet is only a minor element in the development of nutritional osteomalacia.[5] They found absolutely no correlation between the amount of vitamin D in the diet and the amount of calcium, the bone hardener, in the blood. Instead, the elderly with limited sunlight exposure—those who were housebound—showed very low levels of serum calcium even when their vitamin D intake from dietary sources was high. Thus, while it is necessary for the elderly to receive proper nutrition if they are to maintain good health, it is equally essential that they include some sunlight in their regimen. For the housebound elderly, it is important that they get some assistance in getting out of doors occasionally.

For some, getting enough sunshine to insure healthy bones presents even greater problems. Though only a small amount of sun is needed to initiate vitamin D synthesis in the skin, those who live near the Arctic Circle are deprived of even this minimum amount for many months of the year. And as expected, weak bones are a common problem among this population.

For others, sunlight deprivation is a cultural matter. Even people in sun-rich areas of the world can suffer from rickets if they avoid the sun. In Arab countries, for example, rickets is quite common among females who traditionally cover all but a small area around their eyes. And in Kashmir the rickets problem is similarly a female one, since they rarely go outdoors.

Vitamin D Overdose

For those who are unable to get sufficient sunlight because of climatic or cultural restrictions, or for children who need a greater amount than adults, the laboratory synthesis of vitamin D_2 was received as a godsend. But as previously mentioned, dietary D_2 is a weak and inferior substitute for sunlight D_3. Moreover, supplementing diets with D_2 presents some health risks. Since vitamin D is one of the fat-soluble vitamins, a potential for toxic buildup exists if an overdose of D_2 is taken. Vitamin D overdose can lead to the deposit of calcium on the soft tissues of the heart, kidneys, and muscles. Muscular weakness, joint pains, and various other symptoms have also been reported. In Great Britain and several other European countries, the high incidence of vitamin D toxicity has led to a sharp curtailment in the fortification of foods with the vitamin.

Sunlight, then, is essential for proper calcium metabolism. Vitamin pills and even the vitamin D in our food are inadequate substitutes. Fortunately, the amount of sunshine needed for vitamin D synthesis is easily attained by most of us. Even in midwinter in the northern United States, the sunshine falling on the face during a fifteen-minute walk is all that is needed.

SUNLIGHT, THE EYES, AND INTERNAL FUNCTIONING

Sunlight, particularly at the shorter wavelengths, has a limited ability to penetrate the body. Midrange ultraviolet, the UVB, is absorbed by the outer skin layer, the epidermis, and goes no farther. But all wavelengths can enter the eyes.

The eye is especially sensitive to light in the blue-green region, but it will also respond to wavelengths well into the UV range. We know that UVR can affect vision. If a person is fitted with two different contact lenses, one that allows the transmission of UVR and one that blocks it, the UV-blocked pupil will be larger than the other. Though UVR is invisible, the eye is sensitive to it and the pupil constricts upon exposure. This may explain the abnormal sensitivity some contact lens wearers report when in sunlight. The typical UV-blocking contact lens prevents the eye from constricting as much as it would without the lens, and visible light appears brighter. Squinting, headaches, or alternatively, the wearing of sunglasses are the usual results. While we know that visual processes are influenced by visible light and UVR, new and important research is being conducted to determine the effects of light on another portion of the optic pathway.

Sunlight and Internal Functioning

Sunlight's benefits are not limited to skin and vision—the internal organs, glands, and various physiological processes are also affected. Again, the observation of animals gave scientists information on how sunlight causes changes in the body.

An Austrian histologist, S. von Schumacher, noticed that deer antlers from the Tirolean Hunt Show were much larger during years of abundant sunshine. While von Schumacher believed that direct irradiation of the antlers caused an increase in growth, Fritz Hollwich, a German ophthalmologist, advances a different hypothesis.[6] The eyes, says Hollwich, are the windows of the body, and light's entry through them profoundly alters physiology.

Recall the early interest in the color changes animals exhibit when placed on differently colored surfaces. Mast had already shown that the eyes were involved when he changed a plaice's color by changing the color of the background its head rested on. And Hollwich conclusively demonstrated the eyes' unique role when he noted, using a frog, that color did not change with environmental color when the eyes were sutured shut. If the eyes were involved in this process, was it by means of the same mechanism that resulted in vision? Hollwich found that if the frog's visual system was rendered inoperative but its eyes were left open, the color change still occurred. Something other than vision was involved. To distinguish the visual portion of the optic pathway from the portion that induced color change (and much more, as we will see), Hollwich named the latter the "energetic portion" of the optic pathway.

The Pineal Gland

Light sends signals along the energetic portion of the optic pathway, and these signals ultimately reach the pineal gland. This gland has long been a mystery organ. It is a cone-shaped structure about one centimeter in length located in the middle of the brain between the two hemispheres. Historically, the pineal gland was given a more prominent function than in recent times, and René Descartes, seventeenth-century philosopher and scientist, declared that the pineal was the seat of the soul and that it received visual information from the eyes. But in modern times, until very recently, the pineal's only function was thought to be the inhibition of ovarian functioning in females and testosterone secretion in males.

Among lower animals, the pineal has long been known as an important organ. Early in the course of evolution, this body developed the capacity to transmit light, heat, and vibrations. The pineal developed long before the eye and performed some of the same functions. The lizard's pineal even looks like an eye, protruding from the skull and possessing a miniature cornea, lens, and retina.

The pineal's activity, we now know, is regulated by light. In humans as in most other animals, it is active in the dark—the onset of light inhibits it.

The human pineal's sensitivity to light was only recently discovered. Light reaching the retina causes neural signals to be sent to the pineal gland, which responds by altering the amount of the pineal hormone, melatonin, which is synthesized. Melatonin is synthesized in greater quantities under conditions of darkness. (Remember this point: Light *decreases* pineal activity and thus the production of melatonin.)

Melatonin has far-reaching effects on the body. The hormone reaches the midbrain and the pituitary gland via the bloodstream and is associated with modified brain waves (EEG) and feelings of sleepiness. Sexual functioning is also influenced—pineal activity slows down sexual maturation. Rats kept under constant illumination become sexually mature earlier due to lowered melatonin concentrations.

Melatonin rises and falls with a twenty-four-hour, or circadian, cycle. In humans, the level of melatonin peaks at night between the hours of one and three A.M. and is at its low point at about two o'clock in the afternoon.

The Role of Melatonin: Recent Research

The role of the pineal gland and melatonin in human functioning, a topic so long neglected, is currently the subject of intensive investigation, and scientific symposia have been gathered for the exclusive study of this issue.

Psychiatrist Alfred Lewy was the first to discover that light decreases melatonin secretion in humans.[7] While this had long been known to occur in lower animals, a similar demonstration in humans had been wanting. Lewy discovered that humans had been the apparent exception to the rule simply because the intensities of light used in human research were too low. While rats exhibit a decrease in melatonin production with the onset of relatively weak lights, humans require over 1000 times the intensity to produce the effect—2,500 lux or the intensity next to a window on a sunny day. The practical implications of this finding are tremendous—a normal melatonin rhythm requires sunlight or artificial lights of much greater intensity than those in our homes or offices.

Studies of blind persons with no light perception reveal that the majority display unusual melatonin rhythms, either beginning secretion atypically early in the evening or continuing secretion on into the day.[8]

L. Wetterberg has implicated melatonin in what travelers experience as jet lag.[9] People who cross at least two or three time zones, especially when they do it rapidly in an airplane, frequently complain of inexplicable fatigue and loss of alertness which lasts for days. Jet lag does not occur when one flies long distances north-south—that is, when no time zones are crossed.

Wetterberg measured the melatonin rhythms of a family— parents and children—flying from Stockholm to Los Angeles with a time difference of nine hours. The normal circadian rhythms, synchronized with the day-night cycle, were out of phase when the family arrived in Los Angeles. That is, the levels were now higher in the afternoon and lower in the early hours of the morning. Melatonin rhythms, then, while responsive to the day-night cycle, are also governed by internal signals, and the rhythms persist for a while even when day and night are practically reversed. Not until a week had passed did the parents' melatonin rhythms again correspond to the new day-night regimen, shifting by about one hour each day. Interestingly, the twelve- and thirteen-year-old children readjusted more rapidly, reestablishing their new cycles within three days.

The adrenal hormone cortisol also rises and falls with the melatonin cycle, and while it is not yet known whether melatonin actually causes cortisol fluctuation, it is tempting to speculate on some other potential effects of sunlight. Cortisol has a number of effects on the body, and high levels of it are associated with hyperglycemia, a symptom of diabetes. Whether diabetes can be caused by light deprivation is unknown, but H.-J. Jendralski observed a patient in 1951 who had cataracts on both eyes and who was suffering from diabetes insipidus, a condition characterized by excessive excretion of urine and caused by a pituitary disturbance. Following successful removal of the man's opaque lenses, light again entered both eyes freely and all diabetic symptoms disappeared without further therapy.

High levels of cortisol are also associated with decreased antibody formation and thus lowered immunity. This link may explain the numerous observations relating sunlight, or where lacking, artificial UVR, to increased resistance to colds and other infectious diseases. The prevalence of these diseases in wintertime may also be due to decreased levels of sunlight, since melatonin peaks in winter. Whether cortisol shows the same pattern is unknown.

If a normal melatonin cycle is governed by light intensities approaching that of sunlight, what effect does this have on those of us who work indoors? To date, no research has been done to compare the rhythms of indoor and outdoor workers, but psychiatrist N. Okudaira and associates have conducted a naturalistic study of actual light exposure for ten subjects.[10] Seven medical research workers and three housewives wore photometers which measured illumination levels throughout the day and night. The illumination pattern that subjects experienced was radically different from that of plants and animals in natural surroundings, and subjects were exposed to daylight during only brief and scattered periods in the course of a day.

Okudaira notes that melatonin and other human circadian rhythms are not properly synchronized unless there is exposure to bright light for several hours in the course of a day. No artificial light that we encounter in our work places approaches this intensity, and proximity to a window or being outside is required. Failing this, there is a distinct possiblity that people will experience insomnia, depression, and other symptoms of faulty biological rhythms.

SUNLIGHT AND BLOOD CHEMISTRY

The composition of our blood is altered by sunlight in two ways. First, as we have just seen, light entering the eyes pro-

duces systemwide changes in the body. Second, sunlight
striking the skin penetrates to a depth sufficient to irradiate
the bloodstream.

Sunlight need not travel very far into the body to have a
profound impact on the blood. Though the heart is located in
the interior, huge quantities of blood are always in the skin.
Located between the epidermis and the dermis, the extensive
network of capillaries carries about 10 percent of our blood at
any given time. This is twenty or thirty times the amount that
the skin requires to carry on its activities, but the large volume is necessary for the regulation of body temperature. As
blood courses near the body's surface, heat is radiated through
the skin and off into the air, provided that air temperature is
lower than body temperature.

Wavelengths as short as UVA readily pass through the epidermis and irradiate the blood. And in the time it takes to
receive one MED of sunlight—fifteen or twenty minutes on
a summer day—twice the body's entire volume of blood has
passed through the skin. We can expect, then, that sunlight
will have some profound effects on the various constituents of
the blood.

For example, red blood cells, or erythrocytes, which serve
the vital function of transporting oxygen throughout the body,
fall in number when animals are kept in darkness. It is among
the injured and ailing that the effect of light is most crucial.
When, in one study, white mice were made anemic by drain-
ing off 50 percent of their blood volume, those that were later
kept in sunlight regained their original red-blood-cell count,
as well as their health, in two weeks.[11] Artificial enrichment
of sunlight with the addition of UV lights shortened the recovery period to nine or ten days. Those kept in darkness, however, died within two days. Similarly, children who have recently been sick reestablish normal red-blood-cell count more
quickly when exposed to sunshine. Blood hemoglobin (found
in the red blood cells) is also higher in spring and summer
months and after mountain vacations, presumably because of
the more intense solar insolation.

The total level of blood protein—essential to growth, reproduction, and control of cells as well as playing a major role in repairing the body—varies with the amount of sunlight, showing a maximum during the day and reaching a minimum at night. Analyses done on samples collected over a period of eighteen years by blood banks revealed that protein values are lowest in the winter and highest in summer.

The white blood cells, or leukocytes, function mainly to protect the body against disease-causing microorganisms. Their number is also related to the level of illumination entering the eyes as well as that striking the skin. The blind show depressed levels of white blood cells, and the daily rise and fall is quite different from that of sighted people. After prolonged irradiation with either UV or visible light, the white-cell count increases dramatically in sighted persons, quite unlike the more constant, lower level in the blind.

The blood platelets, formed in the giant cells of the bone marrow, are essential to the blood-clotting process. As with so many other constituents of the blood, light affects their number. Rats kept in darkness have depressed platelet levels, while those in subdued daylight exhibit higher levels. Those kept in direct sunlight have the highest numbers. Blind humans also show depressed platelet levels. Hollwich examined a sixty-seven-year-old patient who had lost the sight of one eye in a childhood accident and just recently had lost the sight of the other due to infection. Her platelet values immediately dropped and reached levels typical of the chronically blind within two weeks. Hollwich also notes that patients undergoing successful cataract surgery show a rebounding of platelet levels when light can again pass to the interior of the eye.[12]

Similarly, Hollwich notes that blood cholesterol levels are higher among the blind. But after corrective surgery, levels again fall to normal.

The list of blood chemicals affected by light is extensive and includes calcium, potassium, sodium, and phosphorus in addition to the ones already mentioned. The effects of light on vital processes is indeed profound.

SUNLIGHT AND THE INTERNAL ORGANS:
LIVER AND HEART

Sunlight helps several of the internal organs do their work. The liver, which among other things breaks down toxins in the body so they can be excreted, cannot work at its best without light. When white mice, for example, were given a powerful toxin, the death rate was strongly correlated with the amount of light the animals received.[13] Mice kept in the dark showed the highest incidence of lethality, while those kept in daylight were the most likely to survive. The liver function of the former group was depressed. In an interesting twist, when a nontoxic substance which the liver transforms into a toxic compound was given to mice, the compound was most lethal for animals whose cages were in daylight and least lethal for those kept in the dark. The reduced liver metabolism of the unlighted animals resulted in decreased conversion of the nontoxic substance to a poison and thus led to the reduced death rate. In the next section, where the use of sunlight to treat disease is discussed, you will see how the augmentation of liver function is used to treat certain toxic conditions in humans.

It is interesting to speculate on the role of sunlight in the liver's detoxification of alcohol. Though alcohol experts tell us that the average person metabolizes about one ounce of alcohol per hour, whether waking or sleeping, indoors or outdoors, what we know about the factors that influence the liver's detoxification of substances would suggest otherwise. Though there is no direct evidence of this, it is possible that drinking in the sun results in a lower level of intoxication than would occur when drinking in poorer illumination. More importantly, perhaps, the effect of sunlight on alcohol metabolism may account for an often-noticed but little-understood phenomenon, namely that hangovers seem more intense if drinkers retire immediately after consuming their last drink. Staying awake for a while, allowing light (even artificial light) to stimulate liver function, might reduce the heavy imbiber's hangover. Failing this, getting up and going outside the following day could reduce whatever unpleasantness one incurs through overindulgence.

The heart also benefits from sunshine. Following exposure to sunlight, expecially the UV portion of it, blood pressure can fall by as much as six to eight points. Though the mechanism for this is not completely understood, it is probably due to a dilation of the skin's capillaries which decreases resistance to blood flow and reduces the load on the heart. This dilation also accounts for the reddening of the skin with exposure to the sun.

Our bodies, then, require not only the food, air, and water that our environments provide, but the sunlight as well. The endocrine glands, blood, and internal organs function optimally when there is moderate exposure to sunlight. And while the direct action of sun on skin is an important aspect of well-being, the indirect effect of sunlight as it enters our eyes and sends neural messages throughout the body also merits our attention.

Fritz Hollwich, who has done much to demonstrate the benefits of light entering the eyes, has had an impact on the labor laws of his native Germany. In evaluating the degree of disability caused by work-related blindness, the West German government considers not only the degree of visual impairment, but looks for possible hormonal and metabolic disturbances that can result from blindness as well. If tests of blood chemistry show that the blindness, and thus the loss of photostimulus, has caused endocrine and metabolic abnormalities, the severity of the disability is presumed to be greater and additional compensation is given the worker.

In addition, Hollwich is a firm believer in the use of sunlight to prevent disease and disability. He urges schools, for example, to hold classes out of doors occasionally, weather permitting.

SUNLIGHT AND THE TREATMENT OF ILLNESS

While sunlight can help us maintain optimum levels of health, it can also heal us when we become diseased. As discussed previously, sunlight can aid recovery of red blood cells when an injury has occurred. And the sun can help the liver in its efforts to rid our bodies of dangerous toxins. This effect on liver functioning is of vital importance to the one in six infants born with hyperbilirubinism, or neonatal jaundice. Babies of mothers with incompatible Rh factors often form too much of a substance called bilirubin. An excess can cause mental retardation, cerebral palsy, and even death. The problem is that the infant's liver, especially if he or she is premature, is not sufficiently developed to fully detoxify the excessive bilirubin.

Quite some time ago, hospital personnel noted that infants placed by an open window were able to recover more quickly from neonatal jaundice. In India midwives have used the same practice for ages, placing unclothed infants in the sunlight to cure them. Today, hospitals routinely use three or four days' exposure to light in the treatment of hyperbilirubinism. Light destroys the bilirubin and probably augments the liver's ability to metabolize it as well. While UVR was originally preferred for this purpose, visible blue light has been found to be just as effective and, since it is safe, is most commonly used today.

Older people can also suffer from higher than normal levels of bilirubin, though it does not pose as serious a problem as with infants. And frequently, high bilirubin levels are associated with cataracts. Normal levels are again achieved following successful cataract surgery.

The Treatment of Skin Diseases

A wide variety of skin diseases are now treated, as they have been historically, with sunlight and artificial light. One such skin ailment, vitiligo, is characterized by patches of white, depigmented skin. This is mainly a cosmetic problem, and patients have had some success in darkening afflicted areas by using sunlight in conjunction with substances called psoralens. Psoralens, members of the furocoumarin family, are naturally occurring chemicals which have been isolated from figs, celery, parsley, limes, cloves, and several other plants.

The technique for tanning the depigmented areas of vitiligo

patients was described in the *Atharva-Veda*, the ancient book of India, in about 1000 B.C. Arabian physicians in Egypt practiced the same therapy in A.D. 1250. In both cases, the psoralen was applied directly to the skin followed by exposure to sunlight.

Today, treatment of vitiligo is accomplished similarly, but the psoralen is now taken orally and, more often than not, the light source is artificial. Physicians discovered that when psoralens were applied directly to the skin, the intensity of UV radiation (the UVA band is most effective) necessary to stimulate repigmentation was dangerously high. Instead, purified psoralens such as 8-methoxypsoralen are given orally, and the whole body is irradiated by small, controlled doses of UVA three or four hours after the drug is ingested. Treatment continues for nine months, two or three times a week.

Some physicians still prefer to use sunlight in the treatment of vitiligo. An oral psoralen is likewise administered, but the patient is instructed to sit in the sun after waiting a few hours for the drug to take effect. The increased pigmentation so induced is long lasting—often for as many as nine months—and frequent reexposure is unnecessary. But psoralen therapy with both sunlight and artificial lights has its critics, and there is some evidence of skin cancer and cataracts among susceptible patients.

For some skin problems, the main idea of heliotherapy is to do damage. When used with skin diseases such as herpes simplex (cold sores) and psoriasis, UVR along with a photosensitizing chemical selectively damages the DNA of the invading, unwanted cells. Psoralens are often given to these patients, making them especially sensitive to sunlight or artificial UVA irradiation.

The Heliotherapy of Psoriasis at the Dead Sea

Psoriasis is a chronic, hereditary skin disease characterized by circumscribed red patches covered with white scales. Since the 1950s, dermatologists from Hadassah University in Jerusalem have been treating psoriasis patients at a resort on the Dead Sea, and in recent years over two thousand sufferers have received the therapy, which consists of alternately soaking in the sea and sitting in the sun.[14] This combination of bathing and sunbathing is called heliobalneotherapy and is a popular form of therapy at seaside and mineral-spring resorts around the world.

At the Dead Sea, patients begin with twice-daily, half-hour exposures to the sun and work up to seven hours a day of sunbathing. Sea-bathing time is initially for ten minutes twice daily, increasing to twenty minutes four times a day. Research has shown that 96 percent of the patients were either clear of psoriasis or else significantly improved during the four-week heliobalneotherapy. Within one month of their return home from the Dead Sea, usually to northern European countries, 45 percent reported that their condition returned (as it usually does) but that it was milder than after their return from hospital treatment.

Of the many patients who also complained of arthritic joint pain upon arriving at the clinic, 70 percent reported complete relief from pain at the end of their treatment period. How long this collateral benefit lasted is not indicated.

If psoriasis is usually treated with the oral administration of a psoralen followed by irradiation with artificial UVA, and if it responds favorably to UVA so that it is not necessary to include any of the more damaging UVB in the radiation spectrum, why, then, expose people to as much as seven hours a day of sunlight? And how is such a high rate of success accomplished at the Dead Sea without using the customary psoralens?

The topography of the Dead Sea offers the answer. Located at 1,292 feet below sea level, the Dead Sea is the lowest point on earth. (The lowest point in the United States is in California's Death Valley, 282 feet below sea level.) The quality of UVR that reaches the earth, as will be seen in Chapter Five, is strongly affected by elevation, and it is in the high mountains that the shortest wavelengths of UVR reaching the planet are recorded. In contrast, at very low elevations radiation in the UVB band is greatly attenuated or even absent. At the Dead Sea, the greatest intensity of UVR is found well into the

UVA range, at 365 nm. By two o'clock in the afternoon, the intensity of radiation in the UVB band is very low and no radiation at 297 nm, the wavelength most efficient in causing sunburn and skin cancer, can be measured after this time. Furthermore, humidity in the area of the sea is high, and this scatters UVR, much of which goes off into space. And since UVB is scattered more than UVA, UVB is again preferentially decreased.

Dr. Willy Avrach, Director of the Dead Sea Psoriasis Center, believes that soaking in the sea makes the use of other chemical photosensitizers unnecessary. The mud found there, he notes, contains a natural photosensitizer. In addition, soaking in water alone has a photosensitizing effect on the skin, since this causes it to absorb more UV rays.

Sunlight and the Treatment of Other Diseases

Perhaps the most unorthodox use of heliotherapy is in the treatment of cancer. John Ott, former Director of the Environmental Health and Light Research Institute in Sarasota, Florida, and an ardent proponent of sunlight's benefits, believes that poor-quality light may be responsible for many cancers.[15] Together with Dr. Jane Wright of Bellevue Medical Center in New York, Ott tried heliotherapy with fifteen cancer patients, type unspecified. Since Ott believes that balanced light—with all wavelengths entering the eye—is most beneficial, subjects were asked to spend as much time as possible in sunlight wearing neither corrective lenses nor sunglasses. Fourteen of the fifteen showed no further advancement of tumor formation and some showed "possible improvement" during the course of the study. Results, then, were encouraging if not definitive.

In a more controlled study, mice from a strain extremely susceptible to tumor formation were kept in various light environments. Some were kept under the usual white fluorescent lights, others under pink fluorescent lights, and a third group in natural daylight conditions. The latter group developed tumors an average of two months later than the mice living under artificial lights.

With the connection between sunlight and certain forms of skin cancer being firmly established, it sounds odd indeed to recommend sunlight as therapy for this type of cancer. Ott, however, does just this, maintaining that the restricted spectrum of light emitted by most artificial sources is responsible for these tumors. He once advised a man who had recurrent skin cancers to begin spending time in the sun without his glasses or anything else between his eyes and the sunlight. After four or five months, his skin appeared normal and he no longer needed the surgery recommended by his doctor.

Please note that even Ott does not suggest baking endlessly in the sun without protection. He typically advises sitting in the shade while receiving one's solar therapy. The idea is to be exposed, not overexposed, to all the wavelengths that reach the earth, none being artificially filtered by sunglasses, eyeglasses, or contact lenses.

Reference was made in the first chapter to two cancer-related studies that point to the possible prophylactic value of sunshine. Four decades ago, scientists suggested that exposure to sunlight could reduce the likelihood of more serious forms of cancer though it might cause basal-cell carcinoma and squamous-cell carcinoma, the two less serious forms of skin cancer. And in 1982 researchers in Australia and England, studying the effects of fluorescent lighting on melanoma development, reported that exposure to sunlight actually decreases the incidence of melanoma for those later exposed to fluorescent lighting in their offices.[16]

Ott's observations, along with the two cancer studies discussed earlier, are most intriguing. Cancer itself is a paradoxical disease, its causes and treatments often being similar. That is, the factors that appear to cause the various cancers—radiation and chemical substances, for example—are also used in its treatment. By extension, sunlight is firmly implicated in the causation of some skin cancers but may prove of therapeutic value in the treatment and cure of these cancers as well as the more virulent varieties.

PROPHYLAXIS: SUNLIGHT AND THE PREVENTION OF DISEASE

If sunlight has value in curing disease, and if it is essential for good health, how are people in cold, northern latitudes to get enough in winter? One solution is artificial lights. In northern Europe, research has shown that exposure to UV rays in the winter increases immunity to disease and improves muscular strength as well. In Germany and the U.S.S.R., underground miners are required by law to have preventive UV therapy.

More research has been done in the Soviet Union than anywhere else on the benefits of UVR, especially among groups who receive severely limited amounts of sunlight. Soviet researchers have noted many health problems among mine workers, those living beyond the Arctic Circle, people working in windowless offices, and children in schools. The deficiency of sunlight suffered by people in these circumstances leads to a syndrome the Russians call "Sunlight starvation." Characterized by disorders of the nervous system, an aggravation of chronic diseases, and a general weakening of the body's defenses, sunlight starvation can be treated with UV lights. Work places and schools frequently provide their workers and students with photaria, rooms where lamps rich in UVR are available.[17]

In one Russian school, UV lights were installed in the classroom. Children received between ⅛ and ¼ MED per day (a modest dose too low to elicit tanning or burning) and exhibited improved appetites, weight gain, and increased resistance to illness. And in the mines, as already mentioned, daily UV doses have resulted in a decreased incidence of black lung disease. The Soviet government is impressed with the results of prophylactic UVR. Xenon lamps are being installed in workshops with high ceilings, and legislation requires that UV sources be available to that part of the population residing in northern latitudes or who are out of the sun for all or most of the day. The U.S.S.R. Ministry of Health prescribes suberythemal doses of UVR (less than one MED per day) for all people deprived of sunlight. Artificial UVR is also used on a large scale with agricultural animals such as hens, swine, and cattle in the U.S.S.R.

In our own country, comparatively little work is being done on the disease-preventing attributes of sunlight. Research that is done is largely with animals. One study of interest involved feeding golden hamsters a high carbohydrate diet containing 60 percent sucrose and exposing the animals to different lighting conditions.[18] Half the subjects were under standard white fluorescent lights and half under full-spectrum fluorescent lights which were designed to more closely approximate the daylight spectrum. The first group had an average of 10.9 teeth with cavities, the second only 2.2. This confirms earlier research with humans, conducted in 1938, which found a relationship between hours of sunshine per year for different locales and the incidence of cavities among children.

HEALTH BENEFITS OF DIFFERENT WAVELENGTHS

Which wavelengths of the solar spectrum are most beneficial? Ultraviolet promotes the synthesis of vitamin D and kills bacteria. Infrared wavelengths provide warmth to aching muscles, and blue light is effective in helping infants detoxify excess bilirubin. In all probability, each physical function dependent on sunlight has a different action spectrum (that is, responds best to a particular wavelength). Since all the visible spectrum is present in sunlight, as well as some infrared and UV radiation, most physical processes dependent on light will respond to the sun.

Through nearly all of human evolution, light reaching the body came from the sun. The discovery of fire changed this somewhat, but humans were unable to reliably turn night to day until the development of the incandescent lamp by Thomas Edison about one hundred years ago. This lamp, while adequate for visual functions, emits most of its energy in the red and infrared wavelengths, the amount in the UV band being negligible. What are the consequences of living in a light environment so very different from that provided by sunlight?

John Ott believes that a fair proportion of human illness, including cancer, results from turning our backs on natural light while depending upon the constricted spectrum offered by artificial lights. Ott's interest in the quality of light was an outgrowth of his earlier vocation as a time-lapse photographer. Ott made a short film for Walt Disney that many may still remember—one of dancing flowers, choreographed by varying the light source and taking individual frames at long intervals. Ott showed conclusively that leaf and stem movement is dependent on the direction from which light strikes them. In further film work, Ott discovered that growth is abnormal unless plants receive the full spectrum of sunlight. The growth of ripening apples, for example, is retarded or completely stopped if they receive light from which the UV wavelengths have been filtered. Similarly, Ott found that animal cells show normal growth only when the full spectrum of light is present. Red light alone, the kind so abundant in our incandescent bulbs, is especially deleterious to normal tissue growth.

Light and the Eyes

It took a minor accident for Ott to realize the full implications that light has for human health. Having broken his glasses, he began to spend time in the shade of a Florida palm tree with nothing between his eyes and the sunlight. Amazing things happened, first to his eyes. He no longer needed strong lenses to see, since vision was much improved. Colds and sore throats decreased in frequency and his arthritis, a problem of long standing, got better. X-rays showed improvement in the hip joint that had been a special problem.

As Ott began recommending similar therapy to his friends, he found that those who removed their glasses experienced much improved health. The key: nothing to block out the full spectrum of sunlight, no glasses of any kind to block the UV rays, and, especially, no dark glasses.

Aldous Huxley, British novelist, has written of a similar experience involving sunlight and his eyes. At sixteen, Huxley was virtually blind and had to read Braille. In desperation he undertook the vision-training program of Dr. W. H. Bates. Bates maintains that poor vision is a problem of muscular tension which causes a deformation of the eyes. The key to improved sight, says Bates, is relaxation, not glasses which, while serving as a crutch and making symptoms easier to live with, do nothing to remedy the cause of poor vision. Optometrics, he thought, is a strange branch of medicine in that it never attempts cures and only prescribes medicine (corrective lenses) of ever-increasing strength.

Huxley's vision improved dramatically as he followed Bates's technique. An important part of his therapy was sunning the eyes. In the Bates technique it is essential that a person learn to have normal reactions to bright sunlight, not tense, vision-destroying ones. To achieve this, one closes the eyes, points the face toward the sun, and moves the eyes rapidly across the sun for a few seconds.

Bates introduced his technique in the 1930s, a time when contemporary literature was indicating that bright sunlight was bad for the eyes, causing cataracts and damage to the retina. Sunglasses, donned previously only by those with acute hypersensitivity to sunlight, became popular and were worn widely. You will see in the next chapter that fear of eye damage except under extraordinary conditions was overstated, but by the time corrective evidence was disseminated, sunglasses were firmly established as fashionable and were worn for cosmetic rather than health-related reasons. And in all likelihood, as people began to wear sunglasses, their sensitivity to sunlight increased, and it became necessary to wear them.

Ott views the sunglasses craze with concern. From talking to the daughter of Albert Schweitzer, Ott learned that cancer had become a serious problem only *after* her father had established his hospital at Lambraéné on the west coast of Africa. As Ott suspected, the inhabitants had coincidentally taken to wearing colored glasses in imitation of the white newcomers. He speculates that the lack of full-spectrum light causes hormonal or chemical imbalance, which makes the development of cancer more probable, a viewpoint examined earlier.

Recently, several manufacturers have developed lights that

more nearly approximate the solar spectrum, the full-spectrum lights. And plastics have been developed which, unlike regular glass, allow UV rays to pass through unfiltered. Obrig Laboratories was the first to design a building using full-spectrum lighting as well as UV-transmitting plastic in their windows. And today, both eyeglasses and contact lenses made of UV-transmitting plastics can be purchased. Even persons who must wear lenses over their eyes can now receive the benefits of full-spectrum light.

Further consequences of living under artificial illumination will be covered in Chapter Six.

GROWTH AND BEHAVIOR UNDER THE SUN

All living things, through the process of evolution, have adapted to the natural alternations between light and darkness and, in most latitudes, the changing quality of sunlight throughout the year. All organisms, from protozoa to Homo sapiens, are affected by the day-night sequence and by the changing seasons.

Physiological functions ranging from sleep to sexuality are influenced by changes in light. And sunlight may well be responsible for the many stereotypes we hold concerning the relationship between temperament and climate—the melancholy Danes or the spirited Italians, for example.

Let's now look at the manner in which sunlight influences several aspects of growth and behavior, including a look at effects on human sexual behavior, emotions, and abnormal behavior as well as the behavior of plants and animals.

Plant and Animal Behavior

We know that plants and lower animals respond to changes in sunlight. Chrysanthemums decide when to bloom in response to the shortening of the days (or, more accurately, the lengthening of the nights). Wheat, on the other hand, will head up only when the days grow longer.

Animals also respond to changes of light. In 1925 W. Rowan

captured some North American finches, birds which fly south in the fall to far-distant quarters. He placed the birds under an artificial light sequence which approximated the length of summer days. Upon release, the birds discontinued their southward migration and began to fly north. Rowan thus demonstrated that it was the shortened period of daylight that determined winter migration, the increasing length of the days that told the birds when to go north.

American biologist John Emlen, in later work, discovered how migrating birds get their sense of direction. Using starlings, he prepared half for springtime migration by increasing their periods of artificial light and half for autumn migration by decreasing their light periods. When placed in chambers that were replicas of the night sky, those prepared with increasing light periods oriented themselves northward in relation to the stars, those with decreasing light periods got set to fly south. Though temperature is commonly believed to motivate bird migration, it is in fact the length of day, or photoperiod, that initiates the behavior and celestial bodies that guide it.

SEXUAL FUNCTIONING OF ANIMALS. That light could have an effect on sexual maturation and functioning was also first noted in animals. In 1934 French physician Jacques Benoit discovered that the sexual maturation of male Peking ducks was inhibited by conditions of constant darkness. Those ducks kept in normal daylight conditions matured more quickly. And more recently Richard Wurtman and an associate found that both male and female rats reach sexual maturity earliest when kept under lights that approximate the sun's spectrum.[19] Rats kept under normal fluorescent lights have smaller gonads. Similarly, rabbit breeding (as if it needs any help) is enhanced by full-spectrum light, depressed by artificial light.

Springlike sunshine enhances the sexual motivation of many animals. As the days grow longer, a sequence of events involving the brain, pituitary gland, and gonads is initiated. Joseph Meites of Michigan State University has shown that longer light periods are responsible for spring fever and the accelerated amorous impulses of animals. As the days get longer the pituitary gland secretes hormones which cause the sex glands to enlarge and increase their output of sex hormones.

Human Sexual Maturation

Similar phenomena have been noted in humans. In most animal species, including humans, the pineal gland is intimately associated with sexual maturation. As noted earlier in this chapter, light coming through the eyes creates impulses which are sent to the pineal gland and decrease its level of functioning. Since the pineal gland sends out melatonin, which retards the development of the gonads, turning the pineal off serves to increase sexual maturation. This was first noted in 1898 when a boy whose pineal had been destroyed by a tumor experienced premature sexual maturity. And from time to time we read of bizarre cases of girls who are only five years old giving birth to babies, or of young boys who have developed the bodies of mature men. These anomalies are frequently due to pineal malfunction.

Albert Cook, surgeon and ethnologist of the Perry North Greenland Expedition, found that in this northern latitude, menstruation was often delayed until after a girl's nineteenth birthday. Furthermore, only one in ten women who had already reached menarche continued to menstruate during the sunless winter months. Long periods of continuing darkness apparently stimulate the pineal to overactivity, and sexual functioning slows down or comes to a standstill.

A. Shakir presented a study of the first menstruation of five thousand girls from Baghdad. He found that they were far more likely to experience their first period during the month of June. They also exhibited a growth spurt during the months of increased sunshine.

Sexual Motivation

Human interest in sex occurs in cycles that correspond to the day-night rhythm as well as to the seasonal changes in sunlight. There is little direct experimental evidence of this, but circumstantial evidence is strong. It is known that rats kept in constant light secrete less melatonin from the pineal gland and consequently exhibit a high sex drive. Female rats can be kept in a state of continuous estrus if light is constant. Human pineals respond the same way to light. Most of the daily melatonin output, as already noted, occurs during the nighttime hours and melatonin concentration peaks between one and three A.M.

The sex-suppressing nature of melatonin has been known since 1973 when S. Pavel and his associates administered extracts taken from human pineals to mice and noted the strongly inhibiting effects on the activity of their gonads. Pineal extracts have even been used to treat hypersexual men who were finding it difficult to keep their sexual activities within the limits of the law.

There is also a strong relationship between light and concentrations of testosterone, the hormone responsible for the sex drive. Studies have shown that the blind have significantly lower levels of testosterone. Light-induced impulses do not reach the pineal and there is an overactive secretion of melatonin. Hollwich and an associate found that testosterone levels were restored to normal following successful cataract surgery.[20]

HUMAN SEXUALITY AND SEASON. In early Greek tradition, the periods of greatest sexual abandon correspond to the spring and summer seasons when, as Hesiod put it, "the goats are fattest, wine is best, women most wanton, and men weakest." Similarly, according to ancient Chinese medical lore, spring is the season that awakens human passions.

The famous Belgian statistician, Lambert Quetelet, showed that in Europe conceptions (as inferred from births) reached a maximum in May due, he believed, to an increase in vitality after the long winter. Dr. Cook of the Perry Expedition noted that the Eskimos were languorous and their passions depressed during the long, dark winter months. Desire returned with a vengeance at the first appearance of the spring sun, courtship and love being near-constant preoccupations for the next few weeks. The majority of children were born nine months later. Numerous observations, even in less extreme latitudes, confirm this relationship between sex and season.

But recent statistics reveal that the correlation between conception and season has become diluted. In the United States the number of pregnancies still increases in the month of June (and not only for the recently married), but there is another surge during the shortest days, during the month of December. In fact the December peak has surpassed the June one. Some have suggested that artificial lighting has altered our photoperiods such that the length of time we are in light no longer bears a relationship to the period between sunrise and sunset. We are in a lighted environment, they note, whether natural or artificial, from waking to retiring. This explanation, however, fails to convince, since artificial light is not intense enough to inhibit the functioning of the human pineal.

A more plausible explanation for the December peak in conception is a cultural one. Sexual behavior is never determined by a single factor, whether it be sunlight, or moonlight, or starlight. Cultural factors always enter in. The winter holiday season between Christmas and New Year's is a likely time for conception because people have time off from work and can turn their attentions to other pastimes.

SUNLIGHT AND THE EMOTIONS

Earlier in the chapter, the high rate of emotional problems among the Eskimo population was mentioned. While alcoholism, suicide, and other emotional problems are not unique to the Eskimos, the lack of sunshine may well be a stress factor, among several others, that leads to breakdowns.

Several researchers have speculated that emotional problems among the Eskimos may be due to a lowering of calcium levels during the winter months. Calcium is absorbed in adequate amounts only when vitamin D is also present in the body. And without sunlight, as already noted, vitamin D levels become abnormally low even though one's diet may be rich in it.

Lowered calcium levels cause nervous excitability which,

in the extreme, leads to severe muscle cramps. When calcium deficiency is less pronounced, nervousness and emotional volatility result. The person becomes irritable and less able to deal with stress.

Arctic Hysteria

The Arctic hysterias, a running amok which occurs most frequently during the long winter, may well represent an emotional breakdown aggravated by calcium deficiency. Robert Perry, during his attempt to attain the North Pole, witnessed and described an incident of one such hysteria: "In 1898, while the Windward was in winter quarters off Cape D'Urville, a married woman was taken with one of these fits in the middle of the night. In a state of perfect nudity she walked the deck of the ship; then, seeking still greater freedom, jumped the rail, onto the frozen snow and ice. It was some time before we missed her; and when she was finally discovered, it was at a distance of a half mile, where she was still pawing and shouting to the best of her abilities. She commenced a wonderful performance of mimicry in which every conceivable cry of local bird and mammal was reproduced. The same woman at other times attempts to walk the ceiling of her igloo; needless to say, she has never succeeded."

Calcium and the Circadian Rhythm

Edward Foulkes, an anthropologist who has spent time in the Arctic, believes that the Eskimos suffer not so much from lowered calcium in the winter months as from an upset calcium rhythm.[21] As with many bodily functions, calcium level is normally on a twenty-four-hour cycle, reaching high and low points each day. As is the case with melatonin, these rhythms are synchronized by the day-night cycle. Problems arise when the cycles are upset, and especially when calcium levels do not rise or fall at all—when they become "free running." During the Arctic winter, circadian rhythms become less pronounced or even disappear as people sleep without regard for the clock. Sleep is prolonged and erratic during these months, and normal twenty-four-hour cycles break down.

As calcium metabolism becomes free running, irritability, depression, and general apathy set in. Thus, the alteration of normal calcium cycles may account for the prevalence of emotional breaks when sunlight is absent.

Of course, everyone north of the Arctic Circle does not run amok in the wintertime. The hardships of the cold, dark winter months are stressful for many, but the stress is manifested differently by various people. Men, among whom the Arctic hysterias are less common, are more likely to express it in alcoholism and suicide or in cardiovascular problems. For Eskimo males, these are more culturally acceptable forms of breakdown than is running amok.

Depression

Melatonin reaches a peak in females during the time of menstruation. Perhaps this internal rhythm accounts for the numerous cases of depression reported during menstruation, since melatonin, when administered to depressed patients, increases the intensity of their emotional state.

Severe and chronic depression, while occasionally related to personal catastrophe or loss, often seems to have no external cause and persists for no apparent reason. With increasing frequency in recent decades, treatment has consisted of antidepressant medication which alters the chemical flavor in the synapses, the gaps between neurons through which electrical impulses of the nervous system must pass.

If the depressed person has some sort of chemical imbalance, it is quite possible that events in the environment can cause it. Melatonin can, when administered to a person, exacerbate an already existing depression. Since melatonin level is normally mediated by the light-dark cycle, it is conceivable that light, or its absence, might have something to do with depression.

In 1815 J. F. Cauvin noted that "The influence of light on the morale of man is very powerful. The physician will prescribe light for the sad and weak. When taken with moderate exercise, it will revive lost courage." And A. Hautrive, writing in 1828, said that "in certain mental diseases such as melancholy, the physician will not neglect the most powerful cure nature offers. We know how the gentle climate and sun of Southern Italy and France contribute to the cure of the spleen of the bored Englishman."

Psychiatrist Alfred Lewy and his colleagues at the National Institute of Mental Health have come across many patients who develop a depression in the winter months, one that persists until spring, at which time there is a remission—until the next winter.[22] Lewy decided to alter the photoperiod of a man who showed this pattern, and treatment consisted of advising the man to go outdoors as much as possible as winter approached. In addition, the short days were artificially extended by turning high-energy, broad-spectrum lights on for three hours in the early morning and late afternoon, thereby extending the man's photoperiod. In only a few days, the depression remitted as though it were springtime.

Following Lewy's lead, Norman Rosenthal and associates carried out a broader study with eleven patients who suffered what they termed seasonal affective disorder (SAD).[23] Patients reported an annual depression with the symptoms usually starting between October and December and remitting in March. Many reported a change for the better during those winters they had traveled south to places like Florida or the Caribbean, with symptoms returning shortly after their arrival home. One patient reported being depressed any time during the year when the sky was overcast for three or four days, what she called the "gray sky syndrome."

Rosenthal et al. put these patients on a lighting regimen designed to increase their photoperiod. Patients were instructed to sit before very bright, sunlight-simulating fluorescent lights, installed in their home, for three hours before dawn and three hours after dusk. All reported an improvement in their symptoms, most often after between three and seven days. When the lights were removed, relapse was common.

Some psychiatrists have suggested broadening the patient population so treated to those who are depressed without exhibiting the seasonal swings. Depressed persons whose circadian rhythms are off could have these rhythms delayed or advanced with lights.

Similar treatment of schizophrenia has not been undertaken, but studies of schizophrenic patients maintained on the antipsychotic drug propranolol show that the drug reduces melatonin levels. Perhaps sunlight could also be used as a therapy with this group to more naturally lower their melatonin.

Other Abnormal Behaviors

Abnormal and antisocial behaviors such as crime and alcoholism have been linked to the presence or lack of sunlight. Ever since Quetelet posited his "thermic law of delinquency," criminologists have noted an association between the spring and summer months and violent crime. Of these crimes, rape shows the most pronounced seasonal variation, with a rapid rise in spring and a peak during June and July. Modern criminologists, however, tend to think of climatic determinants of crime as simplistic and mechanistic and turn more readily to social explanations. They note that crimes such as rape are more probable when doors and windows are left open on hot summer nights or when people are more likely to be outdoors. Nevertheless, social explanations do not rule out climatic ones, and both can exist simultaneously.

Studies on alcoholism tend to focus either on genetic dispostitions (as with Alcoholics Anonymous) or on environmental factors such as stress. An interesting twist on the environmental approach is represented by studies of the relationship between light and drinking. Irving Geller has conducted experiments with rats in which he tried to induce them to drink alcohol instead of water by creating different kinds of stress.[24] But no matter what Geller did, his rats preferred water to alcohol—except during weekends, when they would binge on the ethyl. Geller discovered that the automatic timing device for the laboratory lights was not functioning properly and that the rats were being left in total darkness all weekend. Melatonin levels were higher after the weekends, and Geller wondered if this might be the stressor that induced alcohol consumption. The hunch was confirmed when he injected rats with melatonin and found that they again turned to alcohol without being under any other type of stress. Based on this research, it is possible that human alcoholics either have very active pineals or that they get too little pineal-deactivating light.

Alternatively, alcoholics may be unusually sensitive to normal levels of melatonin. In studies of alcoholic patients, depressed levels of melatonin are invariably reported. Perhaps alcohol keeps the melatonin down to a low and tolerable level. Sunlight could do the same.

CONCLUSION

Sunlight is essential to good health. It helps us maintain optimal conditions within the body, speeds recovery from injury and illness, and can be used preventively to ward off common infectious diseases by improving the body's immunological climate. It even helps us maintain psychological health.

But as with many things, more is not necessarily better than less. Even in the winters in the northern United States, adequate amounts of vitamin D can be synthesized in the skin by taking walks during the lunch hour. When sunlight is virtually absent and artificial light is deemed necessary (as by researchers in the Soviet Union), the doses prescribed are very small. (The doses received in our new tanning salons are decidedly immoderate, however. See Chapter Six.)

But too often the quickening of the body and the euphoria experienced with the first caressing rays of spring encourage us to overindulge in the more hostile summer sun. A good thing turns into a menace as several MEDs are soaked up on a single outing. Too many of us believe that, as there is no such thing as too thin, it is impossible to be too tanned.

If sunlight is considered a drug—and it certainly has some salutary effects on mind and body—then overexposure constitutes drug abuse. And the consequences of this addiction, predictably, are unpleasant.

4

THE SUN AS FOE

Wear protective clothing: broad-brimmed hats, long-sleeved shirts and blouses, and slacks instead of shorts. If possible, carry a parasol.
—Alfred W. Kopf, M.D., 1982

. . . and let them not see the sun.
—Psalms, 58:7

SHADES of the nineteenth century. The first quotation, by a leading authority on skin cancer, conjures images out of the past of ladies protecting every square inch of skin from sunlight with ankle-length dresses and high collars that cover the entire neck. Hoods having never been in fashion in the West, a parasol is angled against the needles of the sun. Short of staying out of sunlight altogether (recommended by Dr. Kopf as the first line of defense), not much more can be done.

Obviously, this advice is hyperbole. But while we are well aware of the benefits sunshine brings to us, we must at the same time recognize its potential for harm. The sun is a benefactor, but one that demands our respect.

It is not at all difficult to find lists of a dozen serious health hazards caused or aggravated by sunlight. In fact, as modern medical science turns its attention to sunlight, it tends to explore the pathology associated with the sun to the relative exclusion of its benefits.

The problems with sunlight are largely human ones. Other animals adapt admirably to the sunshine present in their environments. If it is plentiful, they generally evade it by staying in their burrows, by finding shade, or by coming out only in the morning or evening. Frequently, their body coverings are adapted to this aspect of their environments. Plants are similarly adjusted to the amount of sun available. The thick or waxy covering of desert plants protects more delicate tissue from a surfeit of sunlight.

Humans, through evolutionary processes, also developed skin color, and perhaps eye color, appropriate to their latitudes. If we had all stayed where our ancestors came from, the incidence of sun-related problems would be small. But we are a mobile species, and our distribution throughout the world has less and less to do with skin color as time goes on.

To a large extent, it is the light-skinned members of our species who have problems with sunlight. And sun-induced damage becomes a greater threat to whites living near the equator. Cultural factors also play a role. The predilection to get as much sun as possible, to play or work outdoors with a minimum of clothing, abets the problem.

SUNLIGHT AND THE EYES

In Chapter One I noted some of the fears, especially prevalent in the 1920s, concerning eye damage caused by solar radiation and how these fears launched an enduring cultural phenomenon, the wearing of sunglasses. F. Schanz, an ophthalmologist, reported that the glare of sunlight was sufficient to damage the retina as well as the lens, leading to cataracts. Though he later recanted and even began using UV therapy in the treatment of eye disease, this issue of sun-caused damage has recurred through the years. The best evidence, however, is reassuring, and under normal conditions there is no need for concern about the lens and retina. The tendencies to orient ourselves away from bright sun, to close our eyes, or to blink are about all the protection that is needed. The pupils' involuntary constriction further protects the retina from bright light. These reassurances notwithstanding, Americans still spend $800 million a year on sunglasses.

Most of us wear sunglasses for style or comfort, not for eye protection. It is disconcerting to learn that some sunglasses may actually increase the chances of cataract formation, though not of retinal damage. Sunglasses are designed to cut down on visible light, on brightness, and all dark lenses achieve this goal. Unfortunately, many also have what are called UV windows—they fail to block out UVA. When twenty-five randomly chosen sunglasses were tested, a third of them transmitted more UVA than visible light, and in one model, the UV received by the lens was fully double what would have been received if no sunglasses were worn.[1] The researchers found no relationship between either cost or lens color and UV transmission.

People outfitted with sunglasses of this type are worse off than if they wore nothing, since the decrease in brightness allows the pupils to dilate and reduces squinting, two mechanisms which naturally protect the eyes from light. More UVA impinges on the lens than if the eyes were naked. The wearing of a broad-brimmed hat is much superior to this kind of sunglasses, as it eliminates squinting while cutting down significantly on the amount of UVA entering the eyes.

Sunglasses are now available that are guaranteed to eliminate UVA. Plastic UV-400 lenses are designed to absorb all wavelengths below 400 nm. And SpectraShield Human Lens-II is a glass lens with a reflective coating designed to reflect all UVR below 400 nm as well as all infrared radiation above 700 nm. While infrared radiation from the sun is not powerful enough—unless we look directly at it—to cause eye damage, it can cause discomfort.

Dermatologists sometimes advise their patients to wear sunglasses, but they are less concerned with eye damage and more interested in protecting the sensitive skin of eyelids and the area around the eyes from UVB. Most sunscreens cannot be applied too closely to the eyes without irritation, so shielding the area with dark sunglasses provides protection. Habitual squinting, moreover, can cause wrinkles.

When we are on snow, however, sunglasses do protect the eyes. Keratoconjunctivitis (inflammation of the cornea and conjunctiva) is caused by high levels of UVB. The effects of keratoconjunctivitis (also called "welder's flash") are usually temporary. Sunlight by itself is not intense enought to cause the inflammatory reaction, but problems develop when it is augmented by snow. Snow reflects 85 percent of the UVB and 89 percent of all visible light, approximately doubling the intensity of glare. Moreover, if the eyes must be left open there is no way to escape the light of a snowscape—it comes from above, below, and all around, giving one the sensation of being in a light box. Sunglasses are essential under these conditions.

If most ophthalmologists are not very concerned about eye damage from sunlight under normal (and snowless) conditions, what are we to think of the infrequent but persistent accounts of permanent visual problems caused by sunlight? After World War II, for example, English and Belgian soldiers returning from prisoner-of-war camps in Indonesia and the Far East suffered high rates of *irreversible* scotoma, blind spots within the visual field. POWs incarcerated in Germany also exhibited scotomas, but these were *reversible* upon administration of B-complex vitamins. Since both groups of POWs were on semistarvation diets, the best guess is that the high levels of UVR and visible light in the tropical POW camps contributed to irreversible eye damage, but only when vitamin deficiency was severe.

Similarly, a report published in the *Lancet* revealed that aborigines suffer from cataracts at a far higher rate than do nonaboriginal Australians, and that aborigines living close to the equator have the highest rates.[2] The authors attribute the differences in cataract rates to UV intensity, greater for aborigines because their crude homes offer little protection from sunlight, and greater as UV intensity increases at locations nearer the equator. As with POWs, however, these differences may be due to nutritional factors and perhaps sanitary conditions as well. We can say that most *healthy* people seem to suffer no permanent eye damage from sunlight, though to be cautious, people working outside year-round might consider investing in sunglasses that reflect or absorb UVA. For those spending time in snowscapes, they are a must.

Sun Gazing

We are all aware of the dangers of staring at the sun for even a short time. This practice *will* result in retinal burns, what is called chorioretinal injury. But in this case the burns bear no relation to sunburn and are not caused by UVR. The infrared rays, with their heat, cause actual thermal burns on the retina, leading to a permanent loss of vision. Sometimes the visual loss is localized in the peripheral areas of the retina and may pass unnoticed. But if the damage is to the fovea, the depressed area of the retina where vision is clearest, loss of visual acuity may be severe. Cataracts of the lens can also occur as a result of excessive heat, but the heat of sunlight is not sufficient to cause this.

How does the retina suffer a burn when the heat of the sun is insufficient to cause a thermal burn on the skin, and is even too weak to damage the lens of the eye? The answer is to be found in the focusing processes of the eyes, which gather light (and heat) from a wide area and concentrate it on the retina. Consequently, temperatures on the retina can be much higher than on the skin or in the lens.

EYE INJURIES AND THE SOLAR ECLIPSE. Prior to the solar eclipse of March 7, 1970, the National Society for the Prevention of Blindness carried out a publicity campaign warning of the dangers involved in viewing the eclipse directly. During the previous eclipse, severe cases of eye damage had resulted from looking at the sun through sunglasses, double photographic negatives, smoked glass, cameras, and telescopes as well as with the naked eye. The Society spread the word that the only safe way to view an eclipse is by projecting the sun's rays through a pinhole punched in a piece of cardboard onto a second piece of white cardboard—or by watching it on television. In spite of the campaign, 145 cases of serious eye damage were reported after the 1970 eclipse, 80 of them permanent. Lesions of the macula (within which the fovea lies) were the most commonly reported injuries, and those afflicted developed scotomas in the visual field. Two people suffered total visual impairment.

SUNBURNING AND SKIN DAMAGE

Many of us consider the most common sun injury, sunburn, to be only an annoyance. Minor and infrequent sunburns are just that, but repeated and intense ones may cause serious trouble.

Untanned, white skin reacts to sunlight very quickly. In June only fifteen to twenty minutes of midday sun is sufficient to start the sunburning process. Even with exposure this brief, four reactions can be noted. First, a faint red color becomes visible. This is not true sunburn, which will develop several hours later, but only an immediate and transitory reddening. Second, many people exhibit an immediate pigmentation or darkening. This "quick tanning" occurs if ample pigment is already in the skin, either from natural skin color or the faded pigment left over from earlier tans. Third, the skin emits a characteristically pleasant odor, and fourth, vitamin D synthesis begins. No problems so far. However, if one increases the length of exposure by five to ten times (on average untanned skin), the intense redness, pain, and swelling that most of us are familiar with are in the works. Some four to seven hours after first exposure, blood flow in the skin increases and is responsible for the deep, enduring red coloration. Prickle cells, located in the epidermis, are killed in large numbers and there is a rupturing of lysosomes, protein enzymes which protect the body against infection. Indeed, infection frequently follows a severe sunburn, and medical care is sometimes necessary. As the blood vessels dilate, serum leaks into the skin itself and causes edema, or swelling.

If the person is foolish or unlucky enough to extend exposure beyond this point, a third-degree sunburn develops. The serum leakage becomes intense and blisters form between the dermal and epidermal layers of the skin. Chills, vomiting, and even delirium may occur. Whether the sunburn is second degree (swelling) or third degree (blistering), it is this serum between the layers of the skin that eventually causes the outer layer to peel off.

The remedies that people turn to after receiving a painful burn are many, but it is important to note that relief is only symptomatic and that no treatment is going to reverse the damage done. For relief from pain, some recommend soaking in a bathtub of tepid water with two cups of cider vinegar or bicarbonate of soda added. Michael Schreiber, a Tucson dermatologist, recommends applying Avveno powder (an oatmeal extract) to the skin and taking aspirin to decrease inflammation as well as reduce pain. Within twenty-four hours, most of the recognizable negative reactions to sunburn subside.

Needless to say, a person should not be in the sun for several days after a sunburn. Besides starting the messy business all over again, those who sunbathe with a burn run a risk of developing mottled skin, patches that persist even after tanning—a process initiated right along with the sunburn—has come and gone.

Chronic Overexposure

Occasional sunburns, while avoidable, do not present much of a problem to normal skin. However, people with very fair skin—types I and II— are advised to take precautions to avoid any sunburns.

Chronic overexposure is another story. If a person habitually gets too much sun over a period of years, the following permanent changes in skin can result:

SKIN COLOR. Sun-beaten skin takes on a permanent change in color. Persons with skin types III through V often become darker even in the absence of more sun. Fairer-skinned people may notice their complexions turning pinkish with a yellow underlay. This is due to a thinning of the epidermis, which makes skin more transparent. In addition, repeated sunburns can damage the skin's blood vessels, causing them to become flabby, enlarged, and engorged, imparting a florid color to the skin.

PATCHY TANNING. After experiencing chronic overexposure, future tans may appear splotchy. Excessive sunlight can damage the melanocytes in which the melanin-producing process begins, resulting in skin with tanned areas surrounded by unpigmented areas. Thus, excessive sunlight can damage the very cells that produce a tan.

DRY SKIN. In time, sunlight can also cause damage to the sweat glands and the sebaceous gland cells. The sebaceous glands normally secrete a fatty substance called sebum which lubricates and softens the skin. When damaged, however, less sebum is secreted and skin becomes dry and poorly lubricated. The skin may take on a leathery appearance which leads, in the extreme, to pachyderm, or "elephant skin."

Dry, florid, patchy skin is a serious enough problem. But most people reserve their greatest concern for still another problem—sun-induced wrinkling.

WRINKLING

In 1870 Dr. Josiah Gregg noted that "the old people of New Mexican towns look older than in any other country. There is a local proverb that this region is so healthy that the oldest inhabitants never die; but lean, attenuated, and wrinkled, like Egyptian mummies, dry up and are blown away." Gregg, while attesting to the health and longevity of the inhabitants of sunny New Mexico, inadvertently pointed up one of the hazards of too much sun—wrinkling. Since then, dermatologists have clearly established the connection between premature wrinkling and sunlight. With light-skinned people, this wrinkling can easily occur in middle age or sooner. Some prominent researchers, in fact, think that *all* wrinkling is due more to sunlight than to the passage of time. The wrinkled farmer or sailor, they note, often has smooth, unwrinkled skin on parts of the body covered by clothing. And dark-skinned people, especially blacks, seem to age more slowly than whites. This is quite probably due to the increased protection from sunlight furnished by their ample and well-distributed melanin.

The seeds of wrinkling are sown a decade or more before the wrinkles actually appear. Ultraviolet radiation (UVA in particular) penetrates to the lower dermal areas of the skin and alters protein synthesis in this region. Years later, as sun damage begins to appear on the upper epidermal areas (caused by UVB), the preexisting damage in the dermis allows the resultant wrinkles to pass into the dermal region, resulting in deep, furrowed lines in the skin.

The proteins responsible for smooth, supple skin are called elastin and collagen. Normal elastin fibers are capable of stretching and, after being pulled, returning to their former length like rubber bands. Collagen fibers, on the other hand, resist pulling. They stretch only a few percent of their length and account for the skin's resilience.

Sun-damaged skin is called elastotic and is the result of a fragmentation of elastin fibers and the replacement of some collagen fibers by jellylike material. The skin loses its resilience and returns very slowly to its original state after being stretched. Wrinkling and sagging soon follow. These changes occur most frequently in regions of the body routinely exposed to sunshine and in areas that are required to stretch a good deal—the face, neck, and upper chest areas. A person can quite easily attain the benefits of sunlight on the rest of the body while protecting these wrinkle-prone regions with a sunscreen or protective clothing and hat.

The degradation of elastin and collagen leads to cosmetic changes that many of us would like to delay or avoid. But a far more serious form of sun-induced damage is the skin cancer. First, a look at a precancerous condition.

ACTINIC KERATOSIS

Sometimes referred to as precancer, actinic (or sun-induced) keratosis is a very common skin problem, especially among older persons. Keratoses appear as a scaly thickening of the skin, rough, red patches that develop on sun-exposed areas, usually the face, neck, and hands. Actinic keratoses do not always turn into cancer, but do so with sufficient frequency that treatment is advisable. This usually entails the application of 5-fluorouracil, an anticancer drug, to the tumor. This treatment generally leaves no scar and is very successful.

SKIN CANCER

Each year, between 300,000 and 500,000 new cases of skin cancer are detected in the United States, and the incidence grows annually. With increasing interest in outdoor sports and activities, and with a steady migration of people to the Sunbelt states, more people are getting more sun. In addition, the American form of sun worship is beginning to take its toll. In the decades before the 1970s, when there was less concern with the harmful effects of excessive exposure, many people basked obliviously and without protection in the sun in order to attain the "healthy" and status-conferring tan. Since sun damage is cumulative, the effects have only recently been manifested. Sunbathing in the early years, which causes, at worst, painful burns, can lead to the more serious skin cancers we see in middle age and later.

As you would expect, countries and locales that lie within tropical or subtropical latitudes have a higher incidence of skin cancer. Australia, for example, has the highest rate of this affliction in the world. At the moment, southern Arizona is challenging Australia's dubious honor of being number one. The incidence of skin cancer in Arizona is four times greater than in Minnesota.

Skin type is at least as important as where one lives. Australia leads the world in skin cancer, not only because of its latitude, but also because of the racial heritage of its inhabitants. A large proportion of Australia's settlers were Celts who, you may recall, are skin type I's and are most prone to skin cancer. In general, light-skinned, blue-eyed people who burn easily and rarely tan are prime candidates for any solar-induced maladies, including cancer of the skin.

Skin cancer was once called "Seemanshaut" or "sailors' skin" since it was especially frequent among this group. But any occupation that keeps one outdoors increases the risk. In England during the 1920s, observers noted that fishermen, bargemen, and lightermen had an excess of facial cancer; farmers, gardeners, and graziers had an excess of cancer of the hand. The same observations can be made today: construction workers and police officers, as well as farmers and sailors, have a higher incidence than do office workers, houseworkers, clerks in stores, and factory workers.

In the eighteenth century, skin cancers resulted more from exposure to certain chemicals than to sunlight. Sir Percivall Pott was the first to note the frequent occurrence of it in the scrotal skin of chimney sweeps, scrotal skin being especially sensitive to the tars one came in contact with in that profession. And later, skin cancers began to develop in patients who had been treated with medicines containing arsenic. (In Taiwan, 10 percent of the population over the age of sixty has skin cancer due to the arsenic-contaminated water drunk there.) In addition, pitch, paraffin, oil, and radium were later implicated.

It was not until the present century that skin cancer was conclusively linked to sunlight and UVR. Such a relationship had been suspected for some time, but experimental proof was lacking until 1928, when Dr. George Findlay exposed mice to daily doses of UVR and noticed the first cancerous growths after 217 days. The tumors were preceded by changes in the skin—hair was lost from exposed areas and the skin showed an irregular thickening.

While chemicals and compounds such as arsenic, pitch, and coal tar were the first discovered causes of skin cancer, these substances no longer pose much of a hazard, since their danger is recognized and their use carefully regulated. Today, the most common cause of skin cancer is sunlight.

The vast majority of today's skin cancers occur on unprotected areas of the body, namely the face, ears, neck, and backs of the hands. Cancer of the lip occurs more frequently on the lower one, since it protrudes slightly and is therefore more exposed to sunlight.

A study conducted some years ago revealed that twelve times as many skin cancers occurred on the ears of men as women. The survey was done at a time when men rarely wore their hair over their ears. And as women were more apt to wear shorts, tumors more frequently appeared on their legs than the legs of men. These sex differences are less pronounced today.

A contemporary survey investigated the location of 840 basal-cell carcinomas (the most common skin cancer) and found more than 91 percent of them restricted to the head and neck.

Why Skin Cancers Develop

There are currently two prominent theories of skin cancer development. Those cancers related to sunlight exposure are thought to be caused by either or both of the following mechanisms: (1) mutations which occur as sun-damaged genetic material tries to repair itself, or (2) a change in the body's natural defense, its immune system, which occurs upon exposure to sunlight.

SUNLIGHT AND MUTATION. Skin cancers may be the result of solar-induced mutations in the genetic material, DNA. Deoxyribonucleic acid, or DNA, is the master molecule of life, the one that designs all molecules in all living organisms. DNA is a long, coiled molecule resembling a chain, and the order of its "links" constitutes the genetic code, telling a cell whether to construct itself as part of a fingernail or part of a heart. The traffic of molecules in the body is heavy, and every cell experiences the coming and going of billions of them every minute. It is the DNA that tells each cell which molecules to use and respond to, which to let pass.

DNA absorbs and is thus altered by the same wavelengths of UVR found to be most effective in producing skin cancers in laboratory animals, and some researchers take this as strong evidence that it is indeed UV-induced genetic damage that causes these tumors.

Most of the time, the body can repair damage to DNA. A group of enzymes recognizes the damaged area, cuts through, and removes the damaged portion of the chain while another group of enzymes inserts the proper base, or link, and rejoins the chain. The DNA molecule is as good as new.

The normal human cell is capable of making 80,000 of these repairs each hour. Occasionally, however, there are errors in this process and DNA cells that differ from the originals are produced—the chain gets a wrong link. The DNA begins to send out a different genetic message, one that sometimes results in new and abnormal growths in which multiplication is uncontrolled and progressive—cancers, in other words.

In 1982 at a medical symposium at the Roswell Park Memorial Institute in Buffalo, Francois Dautry reported on a major development in cancer research.[3] Dautry and colleagues have demonstrated that normal cells and cancerous cells differ minutely, only one link in the DNA chain being sufficient to differentiate the two. One molecule, of the six billion in the human cell's set of genes, can set off the proliferation of cells known as cancer. The "target," hit by cancer-causing agents including UVR, may be exceedingly small, and one tiny change in the chemical signals may be enough to start the process.

While this exciting research continues, another group of scientists is pursuing a different approach to the explanation of skin cancer.

SUNLIGHT AND THE IMMUNE SYSTEM. Recently, a surprising discovery was made. It appears that cancerous cells arise regularly in all of us, but that a normally functioning immune system recognizes and destroys them. This is the same system that causes the recipients of heart and kidney transplants, as well as skin transplants, to reject their new organs—their immune system recognizes the foreign tissue and gives it no chance to establish itself. People receiving new organs are thus given medication that suppresses this immune response until the transplant is firmly established.

Similarly, UVR suppresses this reaction in the skin. Studies have shown that mice exposed to UV rays are less likely to reject transplanted tumors.[4] Moreover, patients whose immune system is malfunctioning, or who have been given drugs to suppress the system, show high rates of skin cancer.[5]

Chronic overexposure to sunlight may depress the body's normal reaction to malignancies that arise on the skin. Skin cancer cells contain substances called antigens which the body's immune system recognizes. When antigens are present, an alarm is set off and the T-lymphocytes (a type of white blood cell) in the blood attack and kill the cancerous cell, sometimes calling in other elements of the immune system for reinforcement. The foreign cell has been recognized and annihilated. Recent research by Margaret Kripke and associates at the Cancer Biology Program of the National Cancer Institute has shown that UVR interferes with the normal process of skin cancer elimination by reducing the number of T-lymphocytes essential for its destruction.[6]

It is noteworthy that most skin cancers first develop in the summer months when the skin can suffer its greatest assault from UV rays. Tumors take hold, then, at a time when the skin's immune system is too weak to deal with them, when there is the greatest likelihood of UV-induced depression of cellular immunity.

Skin tumors, like other tumors, can be either benign or malignant. The benign ones, such as actinic keratoses, need to be treated since they may later become malignant, but they do not invade neighboring tissue and seldom threaten life while in the benign stage. The malignant tumors, on the other hand, can invade and destroy normal neighboring tissue. Cells may also break away from the malignant tumors and spread through the body by way of the blood and lymphatic systems. When this occurs, they are said to have metastasized.

There are three types of malignant skin cancer: basal-cell carcinoma, squamous-cell carcinoma, and melanoma.

Basal-Cell Carcinoma

Of the various types of skin cancer, basal-cell carcinoma (also called basal-cell epithelioma) is the least serious and most prevalent. First Lady Nancy Reagan was diagnosed and treated for this type of skin cancer in December 1982. Basal-cell carcinoma accounts for over 50 percent of cancers of all types (and there are over one hundred different kinds of cancer). It grows slowly and is the least aggressive of the skin cancers but must be treated to prevent its spread to underlying bone and muscle.

Basal-cell carcinomas, however, do not metastasize. They usually appear in one of two forms—as pale, waxy, pearly nodules that in time may ulcerate and crust over, or as red, scaly, sharply outlined patches. Surgery or other treatment is indicated, and if properly treated, the cure rate for basal-cell carcinoma approaches 97 percent.

Squamous-Cell Carcinoma

This skin cancer occurs at only one-fourth the rate of basal-cell carcinoma. But in common with the basal-cell variety, squamous-cell carcinoma doubles in rate with each ten degrees of latitude toward the equator. While it may resemble basal-cell carcinoma in appearance, squamous-cell carcinoma is considerably more dangerous. The tumor originates in the prickle-cell layer of the epidermis but usually invades the underlying dermis and occasionally metastasizes. Unlike the basal-cell carcinoma, a squamous-cell carcinoma may bleed spontaneously in its advanced stages.

Squamous-cell carcinoma is a rapidly growing cancer, and treatment is more urgently required than is the case with the more slowly growing basal-cell carcinoma. Fortunately, squamous-cell carcinoma has premalignant stages—one of which is the solar keratosis discussed earlier—that are easily diagnosed and treated. Approximately one in five of these solar keratoses will develop into squamous-cell carcinoma so, as noted, even these relatively harmless skin problems should be treated to avoid possible trouble.

But the solar keratosis is only one skin problem that may announce a future squamous-cell carcinoma. A cutaneous horn (a structure that protrudes from the skin, resembles a horn, and may be straight, curved, or twisted) may also be a forerunner as may any area of the skin that is obviously sun damaged. Although infrequent, a squamous-cell carcinoma will sometimes develop on a chronic scar. Only rarely does this cancer show up on skin that appears normal and healthy.

Melanoma

Around 14,800 cases of malignant melanoma are diagnosed each year in this country. Considered the most dangerous type of skin cancer, melanomas spread quickly and metastasize unless caught in an early stage. Of the 5,000 to 6,000 yearly deaths due to skin cancer, two-thirds are attributed to this particular type. Though they begin in the melanocytes (the cells that produce melanin), melanomas can spread to adjacent areas and enter the bloodstream and lymphatic system, thereby spreading the cancer throughout the body. Since they are not restricted to the skin and nearby tissues, some physicians do not include melanomas in the skin cancer category, preferring to group them with the more virulent cancers.

Melanomas range from a tan color to brown or black, though they may occasionally be blue, white, or red. They frequently, though not always, develop from a preexisting mole, and it is wise to bring any change in a molelike growth to the attention of a physician.

In common with the other skin cancers, melanoma appears most frequently on fair-skinned people. In addition, a family history of melanoma is a more important indicator than is the case with basal-cell and squamous-cell carcinoma. The latter two types, while they may appear to run in families, do so only because one's skin type and thus susceptibility to these skin cancers is inherited.

It is essential that melanoma be caught at an early stage. If the tumor is removed before it spreads, the cure rate is now well over 70 percent. Once it has metastasized, however, cure rates fall substantially.

For all three types of skin cancer, prevention is the best medicine, and sense in the sun is recommended as a precaution. But just how strong is the relationship between sunlight and skin cancer?

Sunlight and Skin Cancer

The skin cancer–sunlight connection has long puzzled researchers. Though it has been recognized for years that UVR is related to skin cancer, the strength of this relationship has been questioned. Skin cancer patients as a group do have more exposure time than nonafflicted people, but fully half have spent only a minimum of time in the sun. This has led some researchers to propose that a few high-intensity exposures rather than lifetime exposure may be the critical factor.

If UVR increases with decreasing latitude, then certain health consequences should be evident. Perhaps the best research on the relationship between latitude and the incidence of skin cancer was done in Australia.[7] This island continent is an excellent location for such a study because of its size (it encompasses a large range of latitudes), and because most of the original settlers were of Celtic origin. While persons native to Australia have darkly pigmented skin well adapted to the high intensity of sunshine, persons of Celtic ancestry are only lightly pigmented. Not only does Australia have the highest rate of skin cancer in the world, there is also a striking increase among Celts at latitudes near the equator (northerly latitudes, in this case). Similar studies in the United States, as noted, have also shown a strong relationship between latitude and rates of skin cancer.

The connection between squamous-cell carcinoma and sunshine is more firmly established than is the case with basal-cell carcinoma. The ratio of basal-cell to squamous-cell carcinoma decreases substantially as one moves south, from ten to one in the northern United States to two or three to one in the South. And, as noted, the squamous-cell variety is typically restricted to sun-damaged skin. While this is also the usual site of basal-cell carcinoma, about 30 percent of these occur on areas of the body that do not receive sun. And most interestingly, researchers have had great difficulty in inducing either basal-cell carcinoma or melanoma by exposing laboratory animals to UVR. The cancers so produced are invariably squamous cell.

Melanomas are more common on sun-exposed areas—women's legs and the chests and backs of men. But their relation to the sun is not nearly as well established as is the case with squamous-cell carcinoma. They may appear, for example, under toenails and fingernails (where UV rays cannot penetrate) as well as in the perianal area. And as will be seen in Chapter Six, the fluorescent lighting of offices has recently been linked to melanoma.

While melanomas appear in areas that the sun never reaches, there is good evidence that sunlight is *one* of the factors causing this cancer. A 1975 study, sponsored by the United States Department of Health, Education, and Welfare, revealed the strong relationship between geographic latitude and melanoma.[8] Among white males and females, the mortality rate from melanoma in the states of Washington and New Hampshire was less than half that in Texas, Florida, and Alabama. Moreover, rates among blacks are only one-seventh that of whites, suggesting that melanization furnishes protection. In Australia, the melanoma rate among nonaboriginal groups is strongly correlated with latitude.

The Melanoma Clinic Cooperative Group, an organization comprising four prominent American universities, found that melanoma is more likely to occur on parts of the body subject to excessive UV exposure and is rarer on the areas normally covered by a bathing suit.[9] As do basal-cell carcinoma and squamous-cell carcinoma, the melanomas occur most frequently on upper face, upper chest, and upper back. The abdominal area of males, which no bathing suit covers, is also a common site. Melanomas in this area are less frequent for females because of the one-piece bathing suit. Melanomas in the extreme lower back area of males are quite rare, while they occur here at a moderate rate on females. This may be due to the wearing of low-cut bikinis.

On the other hand, when we consider only those people who work in offices, melanomas are more frequent on areas *covered* by clothing such as the trunk of the body. It appears that sunlight may actually decrease the vulnerability to fluorescent lighting by stimulating melanin production and thickening the outermost layer of the skin. The evidence for this, however, is preliminary, and further research must be done before we can make this statement with confidence.

We can conservatively state that sunlight aggravates melanoma, but few dermatologists see sunlight as its sole cause. Very intense though brief exposure of unadapted skin to sunlight may be a factor. But chronic exposure may likewise be involved as evidenced by the increasing melanoma rate in southern Arizona. One researcher has also suggested that a "solar circulating factor" is produced in skin exposed to sun, one which is capable of producing melanomas anywhere on the body.

Oncologists have turned to the controlled conditions of the laboratory to specify the precise relationship between sunlight and the different varieties of skin cancer. Robert Freeman has demonstrated conclusively that, at least with the squamous-cell carcinomas, the wavelengths that cause this skin cancer are the same as those that produce sunburning and tanning.[10]

Freeman's procedure is instructive. He exposed albino mice to identical amounts of energy at 300, 310, and 320 nm. Intensities were low. Tumors developed only in the 300 nm group after an average of 458 days. Next, to test the hypothesis that skin cancer and sunburning are caused by the same radiation, he exposed the mice to different amounts of the three wavelengths, intensities being proportional to their sunburning effectiveness. That is, higher doses were given in the longer-wavelength irradiations, since their sunburning power is weaker. Now animals in each group developed tumors at the same rate. Freeman concludes that the cancer-causing effectiveness of UVR is similar to the sunburning effectiveness, since, when doses producing the same degree of sunburn are given, equivalent numbers of skin cancers occur. All tumors were squamous-cell carcinomas. Parenthetically, only one new tumor developed in the exposed mice in the weeks after irradiation had been discontinued. This offers some hope to those of us who may have sunned ourselves incautiously in the past but are now willing to mend our ways. If we change our habits, we might avoid skin cancer in our later years.

Other researchers have attempted to extend the cancer-causing spectrum into the UVA region (320–400 nm). And in one case, UVA was found to cause squamous-cell carcinoma in laboratory animals. This research, however, involved the continuous exposure of hairless mice to UVA over a period of twenty weeks. Since continuous exposure does not occur in the natural environment, no firm conclusions can be drawn from this work. It is known that UVA in very high doses can cause skin cancer, but the intensity required is hundreds or even thousands of times that of cancer-inducing UVB levels.

In summary, sunlight is most firmly linked to squamous-cell carcinoma and least closely associated with melanoma, with the basal-cell carcinomas occupying the middle ground. But no one doubts that sunlight (in connection with a person's skin type) is one factor, perhaps the most important factor, in the genesis of skin cancer. While experimental evidence, especially for the melanomas and basal-cell carcinomas, is weak or lacking, circumstantial evidence is strong.

In the future, other causes of skin cancer will undoubtedly be discovered. But for now, we can at least limit excessive exposure to sunlight.

The Treatment and Management of Skin Cancer

What can be done if you think you have developed a skin cancer? First, it should be brought to the immediate attention of a dermatologist or family physician. Time is especially critical if you think it may be melanoma, since the growth and spread of this malignant tumor is rapid. But the basal-cell and squamous-cell carcinomas also need attention. Because the growth of the latter two is slow and they are not usually life threatening, many people postpone diagnosis and treatment of these tumors. Such inaction can only lead to more extensive intervention later on.

If cancer of the skin is diagnosed, the most common treatment today is surgery, though radiation therapy and immunotherapy are also used. Radiation therapy may be recommended for areas where the least destruction of normal tissue is desired—on the lips, eyelids, ears, and nose, for example. X-rays are directed to the affected site and the cancerous tissue is destroyed. Immunotherapy, though not widely used, involves the chemical stimulation of the body's own defenses to destroy cancer cells. Other drugs, DMSO, for example, have come into use recently and have been applied to affected areas with success.

Surgical treatment has the advantage of yielding a biopsy, or tissue sample, so that the surgeon can be certain that the area is tumor free. When performed by a plastic surgeon, the operation results in minimal scarring. Suture lines are placed along the natural folds of the skin, called Langer lines, allowing the scar to blend in with natural features. Those people who develop thick, ropy keloidal scars after surgery will find the results less cosmetically pleasing, though steroids can be used to stimulate normal scarring.

Developed in the 1930s, a surgical technique that has yielded excellent results is Mohs's chemosurgery. The involved tissue is fixed by means of a zinc chloride paste which immobilizes it without altering the architecture of the skin. There follows a series of progressively deeper incisions and biopsies until the removed tissue tests negative for cancer. Mohs reports a 96 percent cure rate for basal-cell carcinomas that had resisted other methods of treatment.

After treatment for skin cancer, physicians usually want to see the patient after one month, three months, six months, and a year to make sure the treatment was effective. In the vast majority of cases it is.

A small proportion of basal-cell carcinomas, however, do recur in the same location. This is far more common among young women than with patients of different age or sex and can be attributed to three factors: inadequate excision due to placing cosmetic concerns above health; the surgeon's fear of legal repercussions if the scarring is disfiguring; and the tendency of some surgeons to minimize the seriousness of basal-cell carcinoma.

The patient who has had skin cancer can usually live a normal life as long as there is no intentional sunbathing or other types of overexposure. Avoiding these cannot guarantee that there will be no future cancers in other locations, but it substantially reduces the risk and number of future tumors. The advice set forth in Chapter Five is especially important for those who have had skin cancer diagnosed and treated.

SKIN CANCER, PHOTOSENSITIVITY, AND GENETICS

For most of us, the genetic predisposition to skin cancer is significant only in the case of melanoma. For the other varieties, genetics, as far as is known, plays a role only insofar as it determines the type of skin we are born with. Our habits—mainly our tendencies to reasonable sunning or overexposure—are of greater importance.

The six skin types discussed in Chapter Two cover the range from the very sensitive to those who are relatively insensitive to the sun's harmful effects. But some people have genetic predispositions which make them acutely sensitive to the sun.

People born with dominant oculocutaneous hypomelanism (DOH) are very light complected and have unusually light eye color. While not albinos, they are extremely sensitive to the sun, suffer painful burns, and are at greater risk for skin cancer. True albinism, also a genetic trait, is characterized by serious visual defects, ivory-white skin, and an acute sensitivity to sunlight.

The congenital disease erythropoietic photoporphyria causes suffers to be extremely sensitive to UVA radiation in the 400 nm range, and exposure to sunlight causes pronounced reddening and swelling. The disease process results in the release of large amounts of substances called porphyrins into the bloodstream. The porphyrins absorb UVA and even visible light, thus creating the problem.

The most serious difficulties with sunlight are experienced by xeroderma pigmentosum (XP) patients. XPs have a deficiency that prevents the repair of sunlight-induced damage. In normal skin, UV damage to DNA mends itself after we are out of the sun. As noted previously, even in normal people this repair is not foolproof, and with repeated sun damage the system can make mistakes with neoplasms, or cancers, resulting. But with XPs, there is little repair and these individuals can suffer damage after only a single exposure to the sun. Typically, XPs develop multiple skin cancers in their juvenile years and the disease can be fatal in the person's youth. Prevention is the only answer, and oral sunscreens which offer XPs some protection have been developed.

Fortunately, the above-named genetic traits are rare. And with the exception of DOH, the resulting problems are serious enough that sufferers are well aware that they have them and that they must be careful in the sun. But all of us can run into trouble if we ingest certain substances that cause a hypersensitivity to the sun if we develop an allergy to the sun. These dangers are discussed in the next section.

NONHEREDITARY PHOTOSENSITIVITY

A wide variety of substances that we ingest or come in contact with can cause adverse reactions upon exposure to the sun. Prescription drugs, plants, and chemicals in consumer products can, in combination with sunlight, lead to serious problems.

Drugs

Drugs are associated with two types of adverse reactions to sunlight: the phototoxic and the photoallergenic. The first type, the phototoxic reaction, occurs just after a susceptible person takes a photosensitizing drug. Symptoms are experienced upon the first exposure to sunlight while the drug is still in the system and can include rashes and inflammation, a watery discharge, and scales or crusts followed by an itching or burning sensation. With photoallergenic reactions, the second type of adverse response to sunlight, similar symptoms develop when a person first takes a photosensitizing drug, is exposed to sunlight, and then is exposed to sunlight a *second* time. Photoallergenic drugs need to be activated by sunlight. Fair-skinned people suffer more from both types of photosensitivity than do those with darker skins.

The potentially photosensitizing drugs include some antibiotics (most notably tetracycline), Orinase (taken for hypoglycemia), Diabinese (diabetes), chlorothiazide (high blood pressure), and chlorpromazine and Librium (anxiety). In addition, riboflavin (a B vitamin), oral contraceptives, quinine, and some barbiturates and antihistamines have evoked similar reactions to sunlight.

Of all these, chlorpromazine is the most potentially photosensitizing. Reactions to chlorpromazine occur upon the second exposure to sunlight after taking the drug. This major tranquilizer, then, is a photoallergen.

To a lesser extent, photosensitivity is higher than usual among females who take birth control pills as well as among pregnant women. Both pregnancy and the pill create high levels of estrogen (a female hormone) in the body, and this hormone is responsible for the increased sensitivity. In addition, both pregnant women and those taking the pill may develop melasma, or patches of dark, pigmented skin on sun-exposed areas. Estrogen is again implicated.

The list of potentially photosensitizing drugs is long, and I have mentioned only a few. If your skin is type I or II, or if you have a history of photosensitivity to certain drugs, it is wise to ask your physician about the possibility of this side effect when beginning a new prescription.

Consumer Products

The list of consumer products that contain potentially photosensitizing chemicals is also long. Some of the halogenated salicylanides, for example, can cause these reactions. They are found in body creams, acne medications, deodorant soaps, and antifungal preparations. One such chemical, TCSA, was removed from the market after causing 10,000 cases of sun-related dermatitis. The essential oils used in cosmetics or beauty aids, plants from the furocoumarin group used as perfumes or spices, and the coal tars used in shampoos as well as in eczema and psoriasis medications can have a similar effect.

It was once thought that the optical brighteners added to laundry detergents, starches, and softeners were photosensitizing, but later research has found this not be be the case. But eosin, a chemical previously used in lipsticks, was so sensitizing that it is now banned from the market.

Plants

Hypersensitivity to sun can also be caused by contact with certain plants. Most of the offending plants contain furocoumarins, popularly called psoralens. Contact with plants such as figs, parsley, lemons, carrots, wild parsnips, and giant hogweed have been known to cause a rash after exposure to sunlight. The problem is most commonly seen with cannery workers who pack the commercial products on this list, as well as with gardeners, florists, and even bartenders. Children playing in sunny fields where these wild plants grow can also develop a rash.

But, as we have already seen, the photosensitizing nature of these plants has also been used to benefit. Beginning several thousand years ago, plants in the psoralen group were used in India to treat people with patches of white skin (called vitiligo or leucoderma). Today, dermatologists use psoralens along with heliotherapy or phototherapy to help develop pigmentation on these white areas. Treatment of psoriasis is accomplished similarly.

Nonhereditary Porphyrias

The porphyrias are a group of related diseases caused by a disturbance in the metabolism of substances called porphyrins. Some of these conditions cause photosensitivity, and the hereditary disease erythropoietic photoporphyria has been previously noted. But some porphyrias which result in photosensitivity can develop from substances that one consumes.

The connection between porphyrin and sunlight was discovered in 1912 when the German physician Meyer Betz injected a small amount of hematoporphyrin into his arm. His face subsequently became swollen, his breathing labored, and he became extremely sensitive to light, even from light bulbs. The effect lasted for three months and then gradually disappeared.

A number of drugs can cause porphyria. Among them are some sedatives, painkillers, anticonvulsants, estrogen, and antimalarial agents. Fungicides and insecticides have also been implicated. The most common cause, however, is alcohol damage to the liver.

Those who suffer from the photosensitizing porphyrias experience a pronounced stinging and burning of the skin when exposed to sunlight. Sunburn, blistering, and peeling soon follow. Wavelengths longer than UVB are involved, since these reactions occur even when sunlight comes through a window (which filters out everything below 320 nm). Even the sunlight which penetrates thin clothing causes discomfort.

Chlorophyll is also a porphyrin and has been found to act as a photosensitizer when applied directly to cells in the laboratory. Since most plants contain chlorophyll, it is reassuring to know that since chlorophyll is not absorbed by the gut, it cannot enter the bloodstream and reach the skin. Thus we have no problems with it.

VITAMIN D TOXICITY

We know that excessive amounts of artificial vitamin D can lead to calcification of heart, kidneys, and muscles as well as other unpleasant symptoms. Overexposure to sunlight may produce similar symptoms, though this issue is still a controversial one.

When UVR strikes the skin, it reacts with 7-dehydrocholesterol and produces natural vitamin D, or D_3. Some researchers believe that, with the exception of the protective tanning process, the body has no way of controlling this synthesis—it goes on and on without regard for the stockpile of the vitamin already in the body.

Fortunately, dark-skinned people who live in sun-rich parts of the world have a built-in protection—ample UV-absorbing melanin in the skin. The tanning response similarly limits vitamin D synthesis. The potential problem arises for those with fair skin living in southern latitudes, especially for those who do not tan. But the evidence for this is mixed. One researcher has proposed that whites living in the tropics suffer from vitamin D "intoxication." But other investigators have shown that surplus amounts of vitamin D can actually be broken down by sunlight, helping the body maintain its optimum level of the vitamin. Clearly, the issue of solar-induced vitamin D toxicity needs more study. But while the jury is out, the fair-skinned person who tans poorly might see another reason to be careful when in the sun.

SUNLIGHT AND BODY TEMPERATURE

Radiation in the infrared range heats the earth and our bodies, often producing a welcome sensation. But excessive heat puts a strain on our bodies, primarily on the heart, and must be dissipated if heatstroke or even death is to be avoided.

The temperatures we must contend with when we are in the sunshine often cannot be read on a normal thermometer. Air temperature, as reported by the weather bureau, refers to temperature in the shade several feet off the ground. Thus, when we hear that the temperature is 100 degrees Fahrenheit, the temperature in the sun may be as high as 140 to 150 degrees. Air temperatures as high as 136 degrees Fahrenheit have been officially recorded in the United States (at Death Valley, California), but on the same day, ground temperatures probably exceeded 160 degrees. How can our bodies, which like to be kept at 98.6 degrees, stand so much heat?

Actually, our bodies are well adapted to stand these temperatures—and much higher. A dramatic experiment conducted over two hundred years ago by Dr. Charles Blagden, Secretary of the Royal Society in London, illustrates the tolerance we have to heat. With some friends and a dog, Blagden went into a room heated to 260 degrees Fahrenheit and remained there for forty-five minutes. The men and dog (which was kept in a basket to keep its feet from burning) emerged without ill effects. Evaporation—the vaporization of water excreted by the sweat glands—was the key to cooling. Where there was no evaporation, temperatures rose dramatically. A steak the men took with them was thoroughly cooked. A pot of water with oil poured on the surface to prevent evaporation was heated to boiling. And when water was poured on the floor, thus raising the humidity and retarding evaporation, it became impossible to stay in the room.

People working in the sun on a very hot day sweat about a quart and a half of water per hour, on the average. The highest rate ever observed in an individual was one *gallon* per hour. Sweating rates for those relaxing in the shade are far less. But even in the hottest sun, the healthy individual can attain adequate cooling by drinking plenty of water.

A curious problem arises, however, for people in hot environments. Those living or working under extreme conditions of heat tend to drink too little water for optimum cooling even though plenty is available. Thus, they may undergo voluntary dehydration. There have been reports of travelers dying from dehydration while driving across the desert with a good supply of water in the car. Under these extreme conditions of heat stress, the feelings of thirst are not adequate indicators of our bodies' need for water and the person has to *remember* to drink even though no sensation of thirst is experienced.

In the 1940s, a group of researchers from the University of Rochester carried out studies of men living under extreme desert conditions.[11] The information was necessary, since World War II was extending into North Africa and little was known about the functioning of soldiers in this hostile environment. They discovered that the primary mechanism for preventing a dangerous heat rise in the body is the transfer of heat from the deeper parts to the skin by way of the bloodstream. As water loss becomes extreme, the blood turns thick and viscous, moving more sluggishly and transferring heat to the body's surface less efficiently.

The University of Rochester group found that as water equal to 2 percent of body weight was lost, some subjects reported a strong thirst. But thirst did not increase as dehydration became more severe. At 4 percent weight loss, the mouth and throat felt dry and subjects became listless, sleepy, and impatient. At 8 percent dehydration, the tongue became sticky and swollen, speech was difficult, and salivation stopped completely. This is the condition that old desert rats refer to as "cotton mouth." For reasons of safety, the experiments were not carried beyond this point.

From animal studies, we know that death from dehydration occurs when water loss equals about 14 percent of body weight. At this stage, animals can only be saved by their removal from the hot environment and by being allowed to drink freely. When dehydration reaches this level, the heat-transfer system breaks down and an explosive rise in body temperature results. In humans, death occurs when body temperature reaches 106 to 107 degrees.

When possible, we should try to avoid the intense sunshine that puts such a strain on the heat-transfer system. If we do find ourselves working or playing in the midday sun, proper clothing is important. People fitted with loose, light-colored clothing have only half the heat gain in the sun as do people who are naked.

Drinking plenty of water is also essential. We cannot be trained to need less water. In fact, adaptation to desert and tropical conditions involves an increased ability to sweat. Water consumption actually rises with adaptation as one's body becomes more efficient at cooling itself.

If we find ourselves in a hot, water-deficient situation, what can be done? Not much, actually, short of exerting ourselves only in the early or late hours of the day and seeking whatever shade is available. Rationing water, in spite of the lore we have seen in westerns, only hastens the time when the earliest signs of dehydration set in. Our bodies use whatever water they need, and we cannot slow down the rate of water use by slow drinking.

Water can be drunk too quickly, however, in which case it will be lost through urination. Fortunately, when our bodies are under heat stress, we lose less water through the urine because of an antidiuretic hormone which is secreted. So even if we drink more than is needed, the body will store it for later use and release it in urine only after the heat stress passes. If, even under these conditions, urination is more frequent than usual, we should cut back on drinking, since too much is being lost without providing any cooling effect.

Summer in the City

As we have seen, the normal, healthy person is admirably equipped to deal with high temperatures, provided plenty of water is drunk. But there are people—primarily the elderly or diseased—for whom the infrared rays of the sun are deadly.

The heat wave that hit the Midwest in July 1980 was associated with a dramatic increase in the death rate. In Kansas City, for example, where there were temperatures above 102 degrees Fahrenheit for seventeen days straight (and ten days when temperatures hit 108 degrees), deaths from all causes increased 64 percent over those of the previous two Julys.[12] More than half of the July 1980 deaths in the city were attributed to the heat by the medical examiner. Many of these were due to stroke. Other deaths were due to an aggravation of heart disease and brain damage from falling (as heat stroke is often preceded by dizziness).

The most likely to suffer from heat stroke were the elderly, and those over sixty-five accounted for 71 percent of the cases. Poor and nonwhite people—those without access to air conditioning—were also more likely to suffer.

Interestingly, adjacent rural areas showed only a small increase in death rate during the heat wave—perhaps because rural temperatures are likely to be somewhat lower than in the city, or perhaps because cooling evening breezes were more prevalent.

The elderly, then, are at exceptional risk for problems associated with heat. Dr. T. Stephen Jones and associates believe that this is due to decreased efficiency of the heart (which must pump blood to the skin to dissipate the body's heat) and the vascular system of this population. Similarly, they note, efficiency of sweating decreases with age. The elderly must take special precautions to keep exertion to a minimum during heat waves, to drink plenty of fluids, to cool the body with tepid baths, and to seek air-conditioned shelter when possible.

5

SENSE IN THE SUN

There's night and day, brother, both sweet things;
sun, moon, and stars, brother, all sweet things.
                                    —George Borrow, *Lavengro*

ALL THINGS necessary for life are harmful or deadly when done to excess. Heat burns and water suffocates if too plentiful. Overbreathing, or hyperventilation, can reduce the carbon dioxide tension of the body and cause a loss of consciousness when the body "forgets" to breathe. Similarly, while the body—from eyes to blood and bone—needs sunlight, an excess causes skin to degrade and even permanent eye damage if one looks directly at the solar orb. While darkness, if extended, leads to general physical degeneration as well as visual impairment, darkness at intervals is necessary to maintain essential cycles within the body and to allow the visual system to regenerate the chemicals on which it relies.

Having seen what the sun can do for us and to us, we need to develop some guidelines so we can reap its benefits while avoiding harm. To start with, we should neither avoid sunshine paranoiacally nor sunbathe to excess, considering only the immediate cosmetic effects. As with most things, moderation is recommended. But while most of us have a good idea of what moderation in food and drink is, few of us know enough about sunlight to achieve the healthful, golden mean we seek. No matter how good our intentions, too many of us either overindulge or suffer sunlight starvation.

PLANTS, ANIMALS, AND HUMANS

For plants and animals, getting the right amount of sunlight has become a factor in survival. Desert plants, for example, have evolved tissues that protect against the overabundant solar radiation of their habitat. Cacti develop a thick, waxy outer covering as well as numerous spines which cast their tiny shadows on the stem of the plant.

Though cacti would seem to be among the most sun-resistant plants, this protection is built up over a number of years. The giant saguaro cactus, symbol and sentinel of the Southwest, must spend its first twenty-five years or so in the shade of a nursery plant—usually a palo verde or mesquite tree—or else be burned to death. Many cactus growers have learned another lesson about the plants' development of protection from sunlight: resistance builds slowly through the spring months to reach a maximum in summer. Cactophiles who move their collections too abruptly from house to bright sun will be rewarded only with burned-up plants. We are all aware of the human parallels.

Animals use a number of devices to avoid overexposure to sunlight. Hair, scales, feathers, and shells are but a few of their protective mechanisms. In addition, many animals have pigmented skin which protects them against excessive UVR. Some animals receive additional protection by changing their coat color from season to season. Animals also adjust their behavior to light conditions. Many hunt or browse only in the evening, night, or early morning, thus avoiding midday sun. This adaptation is especially common among desert animals, many of which are out and about in the cool hours and then return to burrows or shade during the day. The origin of these habits has little to do with avoiding excessive UVR, though this is an added benefit. The triggering mechanism for such instincts is commonly water loss or heat buildup produced by the infrared portion of the solar spectrum.

Human Evolution Under the Sun

Plants and animals have evolved to a point where they receive optimum levels of sunlight simply because of tissue structure or instinctive behavior. Humans have been around for three or four million years, and we might expect that the evolution of our tissues and habits would help us adapt to the amount of sunshine in our environment, whether a sunny or sunless latitude. And such is the case.

As for habits, most people around the world avoid the midday sun, "mad dogs and Englishmen" being the exception to this rule. But in industrialized nations there must be a considerable number of mad dogs, not to mention Englishmen, since many people can be found at midday voluntarily shedding their clothes and sunbathing. Whatever the tendency for humans to stay out of the hot sun, it must not be too strong.

Humans have also developed skin color that helps them adapt to the sunlight in their environment. In the tropics where sunshine is plentiful, people native to the region are generally dark skinned while in sunlight-poor northern latitudes, skin color is light. For many reasons, these different degrees of pigmentation have survival value. Where sun is scarce and skin pigment nearly lacking in the population, vitamin D synthesis takes place more readily than if skin had more UV-blocking pigment. (As noted previously, it is not just the amount of melanin pigment that furnishes protection from UV rays, but also the evenness of its distribution in the skin.)

In tropical climates heavily pigmented skin protects against the skin damage that penetration of excessive UVR might cause and may serve, in addition, to prevent the synthesis of too much vitamin D. Light skin is also more resistant to frostbite than darker skin and some have even suggested that skin color varies by latitude because it provides camouflage and thus has survival value.

So far so good. Our bodies seem admirably adapted to the solar conditions of our place of origin. The problem is that humans are very mobile. The descendants of northern European stock who move to Australia expose themselves to conditions far different from those their endless line of ancestors adapted to and evolved by. Dark-skinned persons who move to Chicago are in similar violation of their genetic roots. The white person out of his or her element runs an increased risk of skin cancer. The relocated black person is more likely to develop a vitamin D deficiency. As mobile humans, then, we must have an awareness of who we are and where we are and calculate the optimum amount of sunshine we get accordingly.

INDIVIDUAL DIFFERENCES

People differ in a number of ways that have relevance to sun exposure. Skin color is the most obvious, and the six skin types were discussed in Chapter Two. Certain other hereditary factors (discussed in Chapter Four) also enter into any assessment of individual sensitivity. And the person's history of acquired sun-related skin problems, especially skin cancer, is another important element in deciding how much sun is healthful.

Prior History of Skin Cancer

Persons who have had one or more skin cancers diagnosed and treated are a special category of people. While their skin type has some relevance to their chances of developing future cancers, they must exercise more caution in the sun than those of a similar skin type with no such history. By following the guidelines presented later in this chapter, those with a history of skin cancer can gain all the benefits of sunshine. But they are well advised to forgo the dark tan. Unfortunately, these are often the same people who prize a tan, and the exposure necessary to maintain one contributed to the skin cancer. Also, unlike their more fortunate friends of a similar skin type, persons having had skin cancer can expect few benefits from a tan. While a tan will protect against sunburn, the person with a history of skin cancer should realize that a tan may indicate overexposure for him or her.

Dermatologists agree that it is a virtual certainty that continued sunbathing by a skin cancer patient will produce more neoplasms. Sorrel Resnick, a Miami dermatologist, maintains that the skin never forgets any exposure to the sun. When a critical level of exposure has been reached, problems develop. The first skin cancer indicates that the person has accumulated enough sun to produce that one. Further sunbathing will add damage to injury, and new areas of the body will begin to reach and exceed critical levels, with new skin cancers being the predictable result.

It cannot be said, unfortunately, that by following sensible guidelines for sun exposure the skin cancer patient can be certain of avoiding any recurrence of the problem. The only claim to be made is that they will develop fewer new cancers if they follow the recommended practices and more if they continue overexposure.

ASSESSING OUR EXPOSURE

So many factors enter into the picture when we try to be sensible in the sun. As noted previously, our bodies have no UV-sensing mechanisms, so we must rely on other information. We must know how the quality of sunlight varies with season, time of day, latitude, cloud cover, ground cover, and numerous other conditions.

The factors affecting the quality of daylight, especially in the UV portion of the spectrum, are so numerous that most attempts to predict the intensity of sunlight are rough estimates. Changes in the atmosphere make the prediction of sunlight intensity as difficult as weather forecasting. Nevertheless, if our purpose is to guide our behavior in the sun and not make precise quantitative statements about sunlight, we have plenty of information to help us.

Season

There is surprisingly little variation from month to month for light in the visible spectrum. Year round, there is a peak intensity in the blue-green region (450–500 nm) when the sun is at its zenith at midday. The picture is different when we look at the seasonal variation of UVR. Depending on other factors such as latitude and altitude, the amount of UVR ranges from a few times to hundreds of times greater in summer than in winter. For example, in the United Kingdom Midlands (53 degrees north latitude), the amount of UVB is one hundred times more plentiful in summer than in winter. In the northern United States (45 degrees north), June UVB levels are fifteen times the December level. And in the southern United States (32 degrees north), summer's intensity is about five times greater than winter's.

As you can see, the difference between summer and winter UVR increases as one moves from the equator to the poles. Summer and winter do not differ by much near the equator, but at high latitudes the shorter wavelengths of UVB may be virtually absent in winter though considerable in summer.

The UVB wavelengths are most subject to seasonal fluctuation while the longer UV rays, the UVA, show less variation. For years, P. Bener has been taking measurements of UV fluctuations with a photometer he has set up at Davos, Switzerland, a mountainous site noted for its health resorts. Bener discovered that at a wavelength of 300 nm (in the UVB range), the summer-winter intensity differed by a factor of forty, while at 330 nm (in the UVA range), summer radiation was only 50 percent greater than winter's.

Since we know that the shorter wavelengths are most active on the epidermis of the skin, the extreme variation in the intensity of these wavelengths from season to season has important implications. The person residing in a far northern latitude, especially if he or she has fair skin to begin with, greets the warm, UVB-rich rays of May or June with skin that has a minimum amount of poorly distributed melanin. It takes only fifteen or twenty minutes to receive an MED on this virgin skin, and extended exposure may have dangerous and painful consequences.

The person residing in a tropical or subtropical latitude is exposed to more UVR on an annual basis, but the distribution throughout the year is more equal. As he or she steps into the June sunshine, considerable pigment formed in the winter and spring months is already present, and the likelihood of a sunburn is smaller.

OZONE AND SEASON. Because of the variation in the level of atmospheric ozone throughout the year, the maximums and minimums of UVR occur at somewhat different times than expected. Ozone increases between January and April, with April being the month of maximum concentration. It then begins to diminish in May and June, reaching its minimum point during October.

Because ozone absorbs UVR, the maximum UVR reaches the earth, not in June when the sun is highest, but in July because of lower ozone levels. Similarly, the northern hemisphere receives the minimum of UVR not in December but in January due to increasing concentrations of ozone.

Time of Day

Ultraviolet radiation also varies greatly throughout the day. It is far less intense in the early morning and late afternoon because when the sun is at a low angle, light has a wider band of atmosphere to traverse. Molecules in the atmosphere scatter UV rays, sending about half of the scattered radiation away from earth and off into space. So as UVR passes through more atmosphere, more of it is scattered away from us.

In 1934 W. Coblentz and R. Stair formulated a table for the National Bureau of Standards that shows the relationship between the sun's angle and the amount of UVB in daylight. The amount of sunburning radiation when the sun is highest was put at 100 percent. The following table shows the relative percentage of this radiation at various angles from the sun's zenith:

Angle of Sun from Noontime Zenith	Percentage of UVB Radiation at Zenith
0°	100%
5	98
10	94
15	87
20	79

Angle of Sun from Noontime Zenith	Percentage of UVB Radiation at Zenith
25	70
30	59
35	49
40	38
45	28
50	20
55	13
60	7
65	4
70	2

Even as much as 20 degrees from the zenith (between one and two o'clock), the sunburning power of the sun is still at 79 percent of its maximum. Not until it travels 35 degrees from the zenith (around three o'clock) does the burning potential drop below 50 percent of the maximum. At 55 degrees (around four o'clock), the burning power of the sun is greatly attenuated, being only 13 percent of the maximum. And at any angle greater than 65 to 70 degrees (after five o'clock), the strong burning rays are virtually absent. (All times are sun times, not Daylight Savings, and refer to summertime.)

Sunlight also strikes our bodies at different angles throughout the day. If the sun is directly overhead, it strikes mainly the top of the head, shoulders, and feet. At increasing angles from the zenith, it strikes us more on chest or back, and legs. Thus, it is the top portions of the body (the upper parts of head, nose, ears, and feet) that receive the most direct radiation when the sun is highest and which are in greatest need of protection.

Altitude and Latitude

As one gains altitude, the amount of UVR increases considerably. The UVB present in the high mountains can be as much as ten times greater than at a similar latitude in the midwestern United States. The shortest wavelengths vary the most with altitude, and the spectrum reaching the earth is extended into the UVC range at very high altitudes. Whereas at sea level one rarely encounters wavelengths shorter than 296 nm, in the high mountains wavelengths as short as 278 nm may reach the earth. Solco Tromp, in *Medical Biometeorology*, notes that tourists vacationing at altitudes of six thousand feet or more frequently feel a loss of appetite, fatigue, nausea, and insomnia after spending some time in the sun.[1] These symptoms do not appear when staying indoors. Tromp believes that the very short UV rays encountered at these elevations cause a type of intoxication.

A change in latitude works in much the same way as a change in altitude. The amount of UVR increases as one moves toward the equator, the difference between high and low latitudes being most pronounced in the winter months. As a general rule of thumb, moving three hundred miles toward the equator increases the UV intensity as much as gaining a thousand feet in elevation.

Measurement of UVR at different North American latitudes reveals the following: If we set the annual total of UVR for the Hudson Bay region at 1.0, then the Great Lakes region receives a total of 2.0 units of UVR, the central United States 2.5, the southern United States 3.0, and the lower southwestern United States 3.5 relative annual units of UVR.

Clouds

The amount of cloud cover throughout the year affects the level of solar radiation received at a particular location. The yearly total of solar radiation is influenced more by the presence or absence of clouds in summer than in winter, so areas with summer monsoons will have less annual solar radiation than locations of similar latitudes having winter rains.

Clouds have a profound effect on sunlight, but the difference between clear and cloudy days is greater at the longer wavelengths. Ultraviolet radiation is decreased by clouds, but to less of an extent than we might think and in ways that are very complex.

Basically, clouds *may* decrease the total amount of UVB reaching the earth, since water vapor scatters UVR and some of this bounces off into space. When 50 percent of the sky is obscured by clouds, about 75 percent of the UVB still gets through to us. When the sky is completely overcast, about 50 percent of the UVB remains. These figures, however, are very broad guidelines and cannot be applied too generally. Without more information, one could get into serious trouble by merely assuming that clouds offer protection from UV rays, as the next section shows.

GLOBAL RADIATION = DIRECT SUNLIGHT + SKYLIGHT. All the light that reaches the earth is refered to as *global radiation*. It has two components—that coming directly from the sun *(sunlight)* and that reflected from the sky *(skylight)*. In the UV portion of the spectrum, skylight is often more powerful than direct sunlight. This surprises most people because direct sunlight feels so powerful. But as we have seen, what we feel are the long infrared rays and not UVR.

HUMIDITY AND SKYLIGHT. The relative amounts of direct sunlight and skylight are related to the quality of the atmosphere. For example, water vapor in the form of humidity scatters UV rays that would otherwise pass right through our atmosphere and continue out to space. Some of this UVR is scattered down at us as skylight. Therefore, even though dry desert skies may appear to be saturated with sunshine, there is actually a greater proportion of UVR in the skylight of humid areas. While the total amount of UVR reaching the desert is greater than that reaching a humid area, a smaller proportion of desert UVR comes from the sky and a greater proportion directly from the sun. Shade, therefore, offers more protection from UVB in the desert than it does in more humid areas.

CLOUDS AND SKYLIGHT. In the high mountains of Switzerland, Bener has found that UVR from skylight exceeds that coming directly from the sun. And while global UVR may be cut in half by extreme cloudiness, the UVR from the skylight portion of the total is not much affected. When the sky is 100 percent overcast and the clouds are of the low type, the *skylight* UVR is still 85 percent as powerful as on a clear day. If the sky is completely overcast with high clouds, however, the skylight UVR can be 133 percent as powerful as on a clear day (though global radiation will, of course, be reduced). Most interestingly, towering clouds that do not obscure the sun can actually increase *global* radiation over that of a clear day. With this type of incomplete cloudiness, the maximum amount of direct sunlight gets through and the skylight is increased, since the clouds reflect some of the scattered UV rays down toward us. This information on skylight will have further application when the protection from UVR offered by shade is discussed.

THE SUNBURNING ULTRAVIOLET METER

We have now looked at the several variables that help determine the intensity of UVR at a given time and place, and we can all use factors such as season, time of day, latitude, altitude, and cloud cover to predict UV intensity on a given day. But in many locations around the world, the actual measurement of the amount of UVR striking the earth is taken on a day-to-day basis. In 1958, a meter capable of measuring UVR was designed in Australia in order to correlate the UVR at several sites with the incidence of skin cancer.

In this country, Daniel Berger of the Temple University School of Medicine has designed the Sunburning Ultraviolet (SUV) meter, which measures energy between 290 and 320 nm.[2] The sensitivity of the SUV meter closely approximates the sensitivity of the skin to UVR. That is, it is attuned to wavelengths around 297 nm with decreasing sensitivity to longer and shorter wavelengths. Dr. Berger's meter, originally commissioned by the United States Department of Transportation to help determine the effects of the SST on atmospheric ozone and resulting UV intensity, has been in operation since 1973.

The SUV meter reads the amount of sunburning energy over a period of a half hour and prints the results on a tape. Since we know how much UVR it takes to produce a barely noticeable erythema in normal, untanned Caucasian skin, the SUV can tell us how many minutes at each hour of the day it takes to receive enough UVR to burn. The highest dose ever recorded on the SUV meter was nearly five MEDs in a single hour.

ULTRAVIOLET INTENSITY TABLES

Data for eleven U.S. cities are presented on pages 144–54, and by choosing the city with a latitude and altitude closest to your own location, you can assess your exposure to sunburning radiation during each month at any hour of the day. The top figure in the tables refers to the number of minutes before a person with average, untanned skin will begin to burn. The number below in parentheses indicates the number of MEDs received during that hour. For example, a person in Fort Worth, Texas, during the 10 o'clock hour in the month of July can stay in the sun for 26 minutes before starting to burn, and the skin will absorb 2.4 MEDs of sunburning radiation during that hour.

Please note that an entry greater than 60 does not necessarily mean that you can stay in the sun for more than an hour; simply that less than one MED of UVR would be received during that hour. The person outdoors for several hours should add the MEDs for each hour to determine total dose.

Remember that the values in these tables are for average, untanned Caucasian skin—skin types III and IV. The skin of the tanned person responds much like that of the person with the next higher skin type—that is, a tanned type III can take as much sun as the untanned type IV before burning.

Persons with skin types I and II should be very cautious in interpreting the figures in these tables. Studies have shown that the type I person may start burning in as little as one-quarter of the time needed to burn average skin.[3] Type IIs burn in a little over half the time it takes types III or IV. People with these more sensitive skin types should adjust the values in the tables accordingly.

ALBUQUERQUE, NEW MEXICO LATITUDE: 35.0 DEGREES N. ALTITUDE: 4,950 FEET

	7am	8	9	10	11	12pm	1	2	3	4	5
JAN	300+ (0.0)	300+ (0.1)	188 (0.3)	98 (0.6)	73 (0.8)	71 (0.8)	86 (0.7)	133 (0.5)	300+ (0.2)	300+ (0.0)	300+ (0.0)
FEB	300+ (0.0)	300+ (0.2)	113 (0.5)	65 (0.9)	50 (1.2)	47 (1.3)	55 (1.1)	79 (0.8)	158 (0.4)	300+ (0.1)	300+ (0.0)
MAR	300+ (0.1)	125 (0.5)	61 (1.0)	39 (1.5)	33 (1.8)	32 (1.9)	37 (1.6)	53 (1.1)	98 (0.6)	261 (0.2)	300+ (0.0)
APR	171 (0.4)	67 (0.9)	39 (1.5)	28 (2.1)	24 (2.5)	24 (2.5)	28 (2.1)	40 (1.5)	71 (0.8)	171 (0.4)	300+ (0.1)
MAY	100 (0.6)	47 (1.3)	30 (2.0)	23 (2.6)	21 (2.9)	21 (2.8)	25 (2.4)	34 (1.8)	58 (1.0)	122 (0.5)	300+ (0.2)
JUN	84 (0.7)	42 (1.4)	27 (2.2)	21 (2.9)	19 (3.2)	19 (3.1)	22 (2.7)	30 (2.0)	51 (1.2)	97 (0.6)	240 (0.3)
JUL	92 (0.7)	44 (1.4)	28 (2.2)	21 (2.9)	18 (3.3)	18 (3.3)	21 (2.8)	28 (2.1)	45 (1.3)	89 (0.7)	261 (0.2)
AUG	128 (0.5)	55 (1.1)	32 (1.9)	24 (2.6)	20 (3.0)	20 (3.0)	23 (2.6)	31 (1.9)	54 (1.1)	118 (0.5)	300+ (0.2)
SEP	182 (0.3)	70 (0.9)	39 (1.5)	28 (2.1)	24 (2.5)	24 (2.5)	30 (2.0)	43 (1.4)	78 (0.8)	214 (0.3)	300+ (0.1)
OCT	300+ (0.2)	107 (0.6)	55 (1.1)	38 (1.6)	33 (1.8)	34 (1.8)	43 (1.4)	67 (0.9)	150 (0.4)	300+ (0.1)	300+ (0.0)
NOV	300+ (0.1)	214 (0.3)	97 (0.6)	91 (1.0)	52 (1.2)	54 (1.1)	72 (0.8)	128 (0.5)	300+ (0.2)	300+ (0.0)	300+ (0.0)
DEC	300+ (0.0)	300+ (0.1)	167 (0.4)	97 (0.6)	76 (0.8)	77 (0.8)	98 (0.6)	176 (0.3)	300+ (0.1)	300+ (0.0)	300+ (0.0)

DES MOINES, IOWA   LATITUDE: 41.5 DEGREES N.   ALTITUDE: 951 FEET

	7am	8	9	10	11	12pm	1	2	3	4	5
JAN	300+ (0.0)	300+ (0.0)	300+ (0.1)	240 (0.3)	171 (0.4)	162 (0.4)	194 (0.2)	300+ (0.2)	300+ (0.1)	300+ (0.0)	300+ (0.0)
FEB	300+ (0.0)	300+ (0.1)	250 (0.2)	140 (0.4)	102 (0.6)	95 (0.6)	113 (0.5)	171 (0.4)	300+ (0.2)	300+ (0.1)	300+ (0.0)
MAR	300+ (0.1)	250 (0.2)	120 (0.5)	77 (0.8)	63 (1.0)	62 (1.0)	71 (0.8)	100 (0.6)	182 (0.3)	300+ (0.1)	300+ (0.0)
APR	300+ (0.2)	133 (0.5)	74 (0.8)	52 (1.2)	44 (1.4)	43 (1.4)	50 (1.2)	66 (0.9)	113 (0.5)	250 (0.2)	300+ (0.1)
MAY	154 (0.4)	77 (0.8)	51 (1.2)	40 (1.5)	34 (1.7)	34 (1.8)	38 (1.6)	48 (1.2)	75 (0.8)	143 (0.4)	300+ (0.2)
JUN	113 (0.5)	58 (1.0)	39 (1.6)	30 (2.0)	27 (2.2)	28 (2.2)	31 (1.9)	39 (1.5)	57 (1.1)	103 (0.6)	250 (0.2)
JUL	120 (0.5)	59 (1.0)	38 (1.6)	29 (2.1)	26 (2.4)	25 (2.4)	28 (2.2)	36 (1.7)	52 (1.2)	91 (0.7)	222 (0.3)
AUG	182 (0.3)	84 (0.7)	51 (1.2)	37 (1.6)	32 (1.9)	31 (2.0)	34 (1.8)	43 (1.4)	69 (0.9)	136 (0.4)	300+ (0.2)
SEP	261 (0.2)	109 (0.6)	63 (1.0)	45 (1.3)	39 (1.5)	41 (1.5)	48 (1.2)	69 (0.9)	128 (0.5)	300+ (0.2)	300+ (0.0)
OCT	300+ (0.1)	222 (0.3)	109 (0.6)	74 (0.8)	64 (0.9)	66 (0.9)	82 (0.7)	133 (0.5)	300+ (0.2)	300+ (0.1)	300+ (0.0)
NOV	300+ (0.0)	300+ (0.1)	261 (0.2)	158 (0.4)	130 (0.5)	136 (0.4)	182 (0.3)	300+ (0.2)	300+ (0.1)	300+ (0.0)	300+ (0.0)
DEC	300+ (0.0)	300+ (0.0)	300+ (0.1)	273 (0.2)	200 (0.3)	207 (0.3)	273 (0.2)	300+ (0.1)	300+ (0.0)	300+ (0.0)	300+ (0.0)

FORT WORTH, TEXAS LATITUDE: 32.8 DEGREES N. ALTITUDE: 820 FEET

	7am	8	9	10	11	12pm	1	2	3	4	5
JAN	300+ (0.0)	300+ (0.1)	207 (0.3)	122 (0.5)	92 (0.7)	88 (0.7)	105 (0.6)	162 (0.4)	300+ (0.2)	300+ (0.0)	300+ (0.0)
FEB	300+ (0.0)	300+ (0.2)	128 (0.5)	78 (0.8)	61 (1.0)	58 (1.0)	67 (0.9)	98 (0.6)	194 (0.3)	300+ (0.1)	300+ (0.0)
MAR	300+ (0.1)	158 (0.4)	81 (0.7)	51 (1.2)	41 (1.5)	40 (1.5)	46 (1.3)	63 (1.0)	111 (0.5)	286 (0.2)	300+ (0.0)
APR	222 (0.3)	91 (0.7)	50 (1.2)	35 (1.7)	30 (2.0)	30 (2.0)	34 (1.8)	46 (1.3)	80 (0.8)	176 (0.3)	300+ (0.1)
MAY	140 (0.4)	65 (0.9)	40 (1.5)	30 (2.0)	25 (2.4)	25 (2.4)	29 (2.1)	37 (1.6)	59 (1.0)	120 (0.5)	300+ (0.2)
JUN	97 (0.6)	49 (1.2)	31 (1.9)	24 (2.6)	21 (2.9)	21 (2.9)	24 (2.5)	32 (1.9)	49 (1.2)	95 (0.6)	261 (0.2)
JUL	111 (0.5)	53 (1.1)	33 (1.8)	26 (2.4)	23 (2.6)	23 (2.7)	26 (2.3)	34 (1.8)	52 (1.2)	100 (0.6)	286 (0.2)
AUG	146 (0.4)	66 (0.9)	39 (1.5)	29 (2.1)	26 (2.3)	26 (2.3)	30 (2.0)	41 (1.5)	67 (0.9)	140 (0.4)	300+ (0.1)
SEP	222 (0.3)	91 (0.7)	52 (1.2)	38 (1.6)	32 (1.9)	33 (1.8)	39 (1.6)	57 (1.1)	103 (0.6)	261 (0.2)	300+ (0.0)
OCT	300+ (0.1)	143 (0.4)	73 (0.8)	51 (1.2)	43 (1.4)	43 (1.4)	54 (1.1)	82 (0.7)	167 (0.4)	300+ (0.1)	300+ (0.0)
NOV	300+ (0.1)	286 (0.2)	133 (0.5)	88 (0.7)	73 (0.8)	72 (0.8)	91 (0.7)	150 (0.4)	300+ (0.2)	300+ (0.0)	300+ (0.0)
DEC	300+ (0.0)	300+ (0.1)	194 (0.3)	118 (0.5)	95 (0.6)	95 (0.6)	120 (0.5)	200 (0.3)	300+ (0.1)	300+ (0.0)	300+ (0.0)

146

MAUNA LOA, HAWAII   LATITUDE: 19.5 DEGREES N.   ALTITUDE: 11,089 FEET

	7am	8	9	10	11	12pm	1	2	3	4	5
JAN	300+ (0.1)	102 (0.6)	44 (1.4)	28 (2.2)	23 (2.6)	23 (2.6)	28 (2.2)	42 (1.4)	86 (0.7)	300+ (0.2)	300+ (0.0)
FEB	300+ (0.2)	73 (0.8)	34 (1.8)	22 (2.7)	19 (3.2)	19 (3.2)	23 (2.6)	33 (1.8)	62 (1.0)	176 (0.3)	300+ (0.0)
MAR	162 (0.4)	50 (1.2)	26 (2.3)	18 (3.3)	16 (3.8)	17 (3.6)	21 (2.9)	28 (2.1)	50 (1.2)	136 (0.4)	300+ (0.1)
APR	98 (0.6)	38 (1.6)	23 (2.6)	18 (3.3)	17 (3.6)	18 (3.3)	21 (2.8)	30 (2.0)	53 (1.1)	133 (0.5)	300+ (0.1)
MAY	77 (0.8)	33 (1.8)	21 (2.8)	17 (3.6)	16 (3.8)	17 (3.6)	20 (2.9)	28 (2.1)	48 (1.3)	113 (0.5)	300+ (0.1)
JUN	74 (0.8)	33 (1.8)	21 (2.9)	16 (3.7)	16 (3.8)	16 (3.7)	19 (3.1)	26 (2.3)	42 (1.4)	94 (0.6)	300+ (0.2)
JUL	85 (0.7)	36 (1.7)	21 (2.8)	17 (3.6)	16 (3.7)	18 (3.3)	21 (2.8)	28 (2.1)	45 (1.3)	95 (0.6)	300+ (0.2)
AUG	94 (0.6)	37 (1.6)	21 (2.8)	16 (3.8)	15 (4.1)	16 (3.7)	20 (3.0)	27 (2.2)	44 (1.4)	103 (0.6)	300+ (0.1)
SEP	103 (0.6)	39 (1.6)	22 (2.7)	17 (3.5)	16 (3.7)	18 (3.4)	22 (2.7)	32 (1.9)	58 (1.0)	167 (0.4)	300+ (0.1)
OCT	125 (0.5)	46 (1.3)	27 (2.3)	20 (2.9)	20 (3.0)	23 (2.7)	30 (2.0)	47 (1.3)	98 (0.6)	300+ (0.2)	300+ (0.0)
NOV	214 (0.3)	68 (0.9)	36 (1.7)	25 (2.4)	23 (2.6)	25 (2.4)	33 (1.8)	53 (1.1)	125 (0.5)	300+ (0.1)	300+ (0.0)
DEC	300+ (0.2)	98 (0.6)	45 (1.3)	30 (2.0)	25 (2.4)	26 (2.3)	33 (1.8)	53 (1.1)	118 (0.5)	300+ (0.1)	300+ (0.0)

MINNEAPOLIS, MINNESOTA LATITUDE: 44.9 DEGREES N. ALTITUDE: 820 FEET

	7am	8	9	10	11	12pm	1	2	3	4	5
JAN	300+ (0.0)	300+ (0.0)	300+ (0.1)	300+ (0.2)	240 (0.3)	222 (0.3)	273 (0.2)	300+ (0.1)	300+ (0.1)	300+ (0.0)	300+ (0.0)
FEB	300+ (0.0)	300+ (0.1)	300+ (0.2)	176 (0.3)	130 (0.5)	122 (0.5)	143 (0.4)	214 (0.3)	300+ (0.1)	300+ (0.0)	300+ (0.0)
MAR	300+ (0.1)	300+ (0.2)	146 (0.4)	94 (0.6)	75 (0.8)	74 (0.8)	87 (0.7)	122 (0.5)	222 (0.3)	300+ (0.1)	300+ (0.0)
APR	300+ (0.2)	133 (0.5)	79 (0.8)	57 (1.1)	50 (1.2)	50 (1.2)	58 (1.0)	78 (0.8)	130 (0.5)	286 (0.2)	300+ (0.1)
MAY	150 (0.4)	80 (0.8)	51 (1.2)	40 (1.5)	36 (1.7)	36 (1.7)	41 (1.5)	52 (1.2)	79 (0.8)	150 (0.4)	300+ (0.2)
JUN	120 (0.5)	66 (0.9)	42 (1.4)	33 (1.8)	30 (2.0)	30 (2.0)	34 (1.8)	43 (1.4)	63 (1.0)	107 (0.6)	250 (0.2)
JUL	120 (0.5)	64 (0.9)	41 (1.5)	32 (1.9)	28 (2.1)	28 (2.2)	31 (1.9)	38 (1.6)	56 (1.1)	97 (0.6)	214 (0.3)
AUG	182 (0.3)	85 (0.7)	51 (1.2)	38 (1.6)	33 (1.8)	33 (1.8)	36 (1.7)	47 (1.3)	72 (0.8)	143 (0.4)	300+ (0.2)
SEP	300+ (0.2)	128 (0.5)	73 (0.8)	51 (1.2)	45 (1.3)	45 (1.3)	53 (1.1)	73 (0.8)	130 (0.5)	300+ (0.2)	300+ (0.0)
OCT	300+ (0.1)	273 (0.2)	133 (0.5)	90 (0.7)	78 (0.8)	81 (0.7)	102 (0.6)	162 (0.4)	300+ (0.2)	300+ (0.0)	300+ (0.0)
NOV	300+ (0.0)	300+ (0.1)	300+ (0.2)	207 (0.3)	171 (0.4)	188 (0.3)	261 (0.2)	300+ (0.1)	300+ (0.0)	300+ (0.0)	300+ (0.0)
DEC	300+ (0.0)	300+ (0.0)	300+ (0.1)	300+ (0.2)	273 (0.2)	273 (0.2)	300+ (0.2)	300+ (0.1)	300+ (0.0)	300+ (0.0)	300+ (0.0)

OAKLAND, CALIFORNIA   LATITUDE: 37.7 DEGREES N.   ALTITUDE: SEA LEVEL

	7am	8	9	10	11	12pm	1	2	3	4	5
JAN	300+ (0.0)	300+ (0.1)	272 (0.2)	144 (0.4)	109 (0.6)	105 (0.6)	125 (0.5)	200 (0.3)	300+ (0.1)	300+ (0.1)	300+ (0.0)
FEB	300+ (0.0)	300+ (0.1)	167 (0.4)	94 (0.6)	70 (0.9)	66 (0.9)	77 (0.8)	113 (0.5)	222 (0.3)	300+ (0.1)	300+ (0.0)
MAR	300+ (0.1)	207 (0.3)	95 (0.6)	60 (1.0)	48 (1.3)	46 (1.3)	53 (1.1)	74 (0.8)	133 (0.5)	300+ (0.2)	300+ (0.0)
APR	240 (0.3)	98 (0.6)	54 (1.1)	38 (1.6)	32 (1.9)	32 (1.9)	36 (1.7)	49 (1.2)	83 (0.7)	188 (0.3)	300+ (0.1)
MAY	133 (0.5)	63 (1.0)	39 (1.5)	29 (2.1)	25 (2.4)	25 (2.4)	28 (2.1)	38 (1.6)	60 (1.0)	125 (0.5)	300+ (0.2)
JUN	133 (0.5)	65 (0.9)	39 (1.5)	28 (2.1)	24 (2.5)	23 (2.6)	26 (2.4)	32 (1.9)	49 (1.2)	87 (0.7)	250 (0.2)
JUL	150 (0.4)	67 (0.9)	38 (1.6)	27 (2.2)	22 (2.7)	21 (2.8)	24 (2.5)	30 (2.0)	46 (1.3)	87 (0.7)	240 (0.3)
AUG	207 (0.3)	86 (0.7)	47 (1.3)	31 (1.9)	26 (2.4)	24 (2.5)	27 (2.2)	36 (1.7)	57 (1.1)	120 (0.5)	300+ (0.2)
SEP	300+ (0.2)	111 (0.5)	57 (1.1)	38 (1.6)	31 (1.9)	31 (2.0)	36 (1.7)	50 (1.2)	91 (0.7)	240 (0.3)	300+ (0.1)
OCT	300+ (0.1)	176 (0.3)	88 (0.7)	59 (1.0)	49 (1.2)	50 (1.2)	62 (1.0)	98 (0.6)	214 (0.3)	300+ (0.1)	300+ (0.0)
NOV	300+ (0.0)	300+ (0.2)	143 (0.4)	94 (0.6)	77 (0.8)	81 (0.7)	107 (0.6)	188 (0.3)	300+ (0.1)	300+ (0.0)	300+ (0.0)
DEC	300+ (0.0)	300+ (0.1)	261 (0.2)	150 (0.4)	120 (0.5)	120 (0.5)	158 (0.4)	272 (0.2)	300+ (0.1)	300+ (0.0)	300+ (0.0)

PHILADELPHIA, PENNSYLVANIA   LATITUDE: 40.0 DEGREES N.   ALTITUDE: 220 FEET

	7am	8	9	10	11	12pm	1	2	3	4	5
JAN	300+ (0.0)	300+ (0.0)	300+ (0.1)	240 (0.3)	182 (0.3)	176 (0.3)	214 (0.3)	300+ (0.2)	300+ (0.1)	300+ (0.0)	300+ (0.0)
FEB	300+ (0.0)	300+ (0.1)	231 (0.3)	133 (0.5)	103 (0.6)	98 (0.6)	113 (0.5)	171 (0.4)	300+ (0.2)	300+ (0.1)	300+ (0.0)
MAR	300+ (0.1)	240 (0.3)	79 (0.8)	73 (0.8)	62 (1.0)	60 (1.0)	70 (0.9)	100 (0.6)	182 (0.3)	300+ (0.1)	300+ (0.0)
APR	250 (0.2)	105 (0.6)	63 (1.0)	46 (1.3)	41 (1.5)	41 (1.5)	48 (1.3)	67 (0.9)	113 (0.5)	261 (0.2)	300+ (0.1)
MAY	150 (0.4)	77 (0.8)	49 (1.2)	39 (1.6)	34 (1.7)	36 (1.7)	42 (1.4)	57 (1.1)	91 (0.7)	182 (0.3)	300+ (0.1)
JUN	136 (0.4)	70 (0.9)	47 (1.3)	36 (1.7)	32 (1.9)	33 (1.8)	38 (1.6)	50 (1.2)	77 (0.8)	143 (0.4)	300+ (0.2)
JUL	136 (0.4)	70 (0.9)	44 (1.4)	34 (1.8)	30 (2.0)	30 (2.0)	34 (1.8)	44 (1.4)	67 (0.9)	125 (0.5)	300+ (0.2)
AUG	200 (0.3)	88 (0.7)	54 (1.1)	39 (1.5)	35 (1.7)	36 (1.7)	41 (1.5)	54 (1.1)	86 (0.7)	171 (0.4)	300+ (0.1)
SEP	300+ (0.2)	120 (0.5)	68 (0.9)	50 (1.2)	44 (1.4)	45 (1.3)	55 (1.1)	79 (0.8)	140 (0.4)	300+ (0.2)	300+ (0.0)
OCT	300+ (0.1)	207 (0.3)	109 (0.6)	77 (0.8)	67 (0.9)	71 (0.8)	91 (0.7)	150 (0.4)	300+ (0.2)	300+ (0.0)	300+ (0.0)
NOV	300+ (0.0)	300+ (0.1)	214 (0.3)	136 (0.4)	118 (0.5)	125 (0.5)	171 (0.4)	300+ (0.2)	300+ (0.1)	300+ (0.0)	300+ (0.0)
DEC	300+ (0.0)	300+ (0.0)	300+ (0.1)	250 (0.2)	200 (0.3)	207 (0.3)	286 (0.2)	300+ (0.1)	300+ (0.0)	300+ (0.0)	300+ (0.0)

Salt Lake City, Utah   Latitude: 40.8 degrees N.   Altitude: 4,260 feet

	7am	8	9	10	11	12pm	1	2	3	4	5
JAN	300+ (0.0)	300+ (0.0)	300+ (0.1)	250 (0.2)	182 (0.3)	176 (0.3)	214 (0.3)	300+ (0.2)	300+ (0.1)	300+ (0.0)	300+ (0.0)
FEB	300+ (0.0)	300+ (0.1)	214 (0.3)	122 (0.5)	95 (0.6)	87 (0.7)	103 (0.6)	171 (0.4)	300+ (0.2)	300+ (0.1)	300+ (0.0)
MAR	300+ (0.1)	162 (0.4)	85 (0.7)	58 (1.0)	50 (1.2)	51 (1.2)	62 (1.0)	91 (0.7)	171 (0.4)	300+ (0.1)	300+ (0.0)
APR	214 (0.3)	94 (0.6)	53 (1.1)	38 (1.6)	34 (1.8)	36 (1.7)	44 (1.4)	63 (1.0)	115 (0.5)	300+ (0.2)	300+ (0.0)
MAY	146 (0.4)	74 (0.8)	43 (1.4)	32 (1.9)	28 (2.1)	29 (2.1)	33 (1.8)	45 (1.3)	77 (0.8)	162 (0.4)	300+ (0.1)
JUN	103 (0.6)	52 (1.2)	33 (1.8)	25 (2.4)	22 (2.7)	21 (2.8)	24 (2.5)	34 (1.8)	52 (1.2)	98 (0.6)	286 (0.2)
JUL	113 (0.5)	52 (1.1)	31 (1.9)	23 (2.6)	20 (2.9)	20 (3.0)	22 (2.8)	28 (2.2)	43 (1.4)	85 (0.7)	240 (0.3)
AUG	167 (0.4)	71 (0.9)	38 (1.6)	28 (2.2)	24 (2.5)	24 (2.5)	26 (2.3)	34 (1.8)	57 (1.1)	122 (0.5)	300+ (0.1)
SEP	286 (0.2)	103 (0.6)	55 (1.1)	40 (1.5)	34 (1.8)	35 (1.7)	42 (1.4)	59 (1.0)	113 (0.5)	300+ (0.2)	300+ (0.0)
OCT	300+ (0.1)	188 (0.3)	88 (0.7)	57 (1.1)	48 (1.3)	49 (1.2)	64 (0.9)	105 (0.6)	250 (0.2)	300+ (0.1)	300+ (0.0)
NOV	300+ (0.0)	300+ (0.1)	240 (0.3)	136 (0.4)	111 (0.5)	118 (0.5)	158 (0.4)	286 (0.2)	300+ (0.1)	300+ (0.0)	300+ (0.0)
DEC	300+ (0.0)	300+ (0.0)	300+ (0.1)	300+ (0.2)	214 (0.3)	207 (0.3)	273 (0.2)	300+ (0.1)	300+ (0.0)	300+ (0.0)	300+ (0.0)

	7am	8	9	10	11	12pm	1	2	3	4	5
JAN	300+ (0.0)	300+ (0.0)	300+ (0.1)	300+ (0.1)	300+ (0.1)	300+ (0.2)	300+ (0.1)	300+ (0.1)	300+ (0.0)	300+ (0.0)	300+ (0.0)
FEB	300+ (0.0)	300+ (0.0)	300+ (0.1)	300+ (0.2)	200 (0.3)	176 (0.3)	200 (0.3)	300+ (0.2)	300+ (0.1)	300+ (0.0)	300+ (0.0)
MAR	300+ (0.0)	300+ (0.2)	171 (0.4)	103 (0.6)	86 (0.7)	79 (0.8)	97 (0.6)	136 (0.4)	240 (0.3)	300+ (0.1)	300+ (0.0)
APR	300+ (0.1)	207 (0.3)	81 (0.7)	74 (0.8)	63 (1.0)	61 (1.0)	65 (0.9)	88 (0.7)	143 (0.4)	300+ (0.2)	300+ (0.1)
MAY	194 (0.3)	103 (0.6)	61 (1.0)	48 (1.3)	40 (1.5)	40 (1.5)	44 (1.4)	58 (1.0)	92 (0.7)	176 (0.3)	300+ (0.1)
JUN	130 (0.5)	70 (0.9)	45 (1.3)	34 (1.7)	31 (2.0)	29 (2.1)	33 (1.8)	43 (1.4)	61 (1.0)	113 (0.5)	250 (0.2)
JUL	194 (0.3)	98 (0.6)	57 (1.1)	43 (1.4)	36 (1.7)	32 (1.9)	37 (1.6)	45 (1.3)	65 (0.9)	111 (0.5)	261 (0.2)
AUG	300+ (0.2)	140 (0.4)	77 (0.8)	54 (1.1)	43 (1.4)	43 (1.4)	51 (1.2)	65 (0.9)	103 (0.6)	214 (0.3)	300+ (0.1)
SEP	300+ (0.1)	261 (0.2)	125 (0.5)	79 (0.8)	63 (1.0)	63 (1.0)	69 (0.9)	105 (0.6)	193 (0.3)	300+ (0.1)	300+ (0.0)
OCT	300+ (0.0)	300+ (0.2)	200 (0.3)	130 (0.5)	105 (0.6)	103 (0.6)	133 (0.5)	222 (0.3)	300+ (0.1)	300+ (0.0)	300+ (0.0)
NOV	300+ (0.0)	300+ (0.0)	300+ (0.1)	300+ (0.2)	240 (0.3)	261 (0.2)	300+ (0.2)	300+ (0.1)	300+ (0.0)	300+ (0.0)	300+ (0.0)
DEC	300+ (0.0)	300+ (0.0)	300+ (0.1)	300+ (0.1)	300+ (0.1)	300+ (0.1)	300+ (0.1)	300+ (0.0)	300+ (0.0)	300+ (0.0)	300+ (0.0)

	7am	8	9	10	11	12pm	1	2	3	4	5
JAN	300+ (0.0)	300+ (0.1)	167 (0.4)	92 (0.7)	68 (0.9)	63 (1.0)	71 (0.8)	107 (0.6)	214 (0.3)	300+ (0.1)	300+ (0.0)
FEB	300+ (0.0)	261 (0.2)	103 (0.6)	61 (1.0)	47 (1.3)	45 (1.3)	51 (1.2)	73 (0.8)	133 (0.5)	300+ (0.2)	300+ (0.0)
MAR	300+ (0.1)	133 (0.5)	65 (0.9)	42 (1.4)	34 (1.8)	33 (1.8)	39 (1.6)	53 (1.1)	94 (0.6)	240 (0.3)	300+ (0.0)
APR	171 (0.4)	70 (0.9)	39 (1.5)	29 (2.1)	25 (2.4)	25 (2.4)	29 (2.0)	41 (1.5)	67 (0.9)	154 (0.4)	300+ (0.1)
MAY	128 (0.5)	58 (1.0)	36 (1.7)	28 (2.1)	25 (2.4)	26 (2.3)	31 (1.9)	41 (1.5)	67 (0.9)	143 (0.4)	300+ (0.1)
JUN	100 (0.6)	50 (1.2)	33 (1.8)	25 (2.4)	23 (2.6)	24 (2.5)	29 (2.1)	41 (1.5)	70 (0.9)	136 (0.4)	300+ (0.2)
JUL	113 (0.5)	56 (1.1)	36 (1.7)	29 (2.1)	26 (2.3)	27 (2.2)	33 (1.8)	47 (1.3)	67 (0.9)	125 (0.5)	300+ (0.2)
AUG	133 (0.5)	57 (1.1)	36 (1.7)	28 (2.1)	28 (2.2)	27 (2.2)	33 (1.8)	48 (1.3)	74 (0.8)	154 (0.4)	300+ (0.1)
SEP	188 (0.3)	74 (0.8)	42 (1.4)	32 (1.9)	29 (2.1)	31 (2.0)	39 (1.5)	56 (1.1)	98 (0.6)	250 (0.2)	300+ (0.0)
OCT	300+ (0.2)	107 (0.6)	56 (1.1)	39 (1.5)	34 (1.7)	36 (1.7)	45 (1.3)	71 (0.9)	154 (0.4)	300+ (0.1)	300+ (0.0)
NOV	300+ (0.1)	188 (0.3)	90 (0.7)	61 (1.0)	51 (1.2)	53 (1.1)	68 (0.9)	111 (0.5)	273 (0.2)	300+ (0.1)	300+ (0.0)
DEC	300+ (0.0)	300+ (0.2)	140 (0.4)	85 (0.7)	67 (0.9)	67 (0.9)	86 (0.7)	140 (0.4)	300+ (0.2)	300+ (0.0)	300+ (0.0)

TUCSON, ARIZONA LATITUDE: 32.5 DEGREES N. ALTITUDE: 2,630 FEET

	7am	8	9	10	11	12pm	1	2	3	4	5
JAN	300+ (0.0)	300+ (0.2)	130 (0.5)	71 (0.9)	56 (1.1)	58 (1.0)	70 (0.9)	115 (0.5)	286 (0.2)	300+ (0.0)	300+ (0.0)
FEB	300+ (0.1)	200 (0.3)	80 (0.8)	49 (1.2)	38 (1.6)	39 (1.5)	46 (1.3)	73 (0.8)	158 (0.4)	300+ (0.1)	300+ (0.0)
MAR	273 (0.2)	85 (0.7)	43 (1.4)	30 (2.0)	25 (2.4)	25 (2.4)	30 (2.0)	45 (1.3)	96 (0.7)	250 (0.2)	300+ (0.0)
APR	118 (0.5)	48 (1.3)	27 (2.2)	20 (3.0)	18 (3.4)	18 (3.3)	22 (2.8)	32 (1.9)	61 (1.0)	154 (0.4)	300+ (0.1)
MAY	83 (0.7)	38 (1.6)	24 (2.5)	19 (3.2)	17 (3.6)	18 (3.4)	21 (2.9)	29 (2.1)	50 (1.2)	118 (0.5)	300+ (0.1)
JUN	70 (0.9)	34 (1.8)	22 (2.7)	17 (3.6)	16 (3.8)	16 (3.8)	19 (3.2)	25 (2.4)	41 (1.5)	92 (0.7)	286 (0.2)
JUL	84 (0.7)	39 (1.6)	24 (2.5)	18 (3.3)	16 (3.8)	16 (3.7)	19 (3.2)	27 (2.3)	46 (1.3)	103 (0.6)	300+ (0.2)
AUG	113 (0.5)	45 (1.3)	27 (2.2)	20 (3.0)	17 (3.5)	17 (3.4)	22 (2.7)	34 (1.8)	62 (1.0)	143 (0.4)	300+ (0.1)
SEP	143 (0.4)	55 (1.1)	30 (2.0)	22 (2.7)	19 (3.1)	20 (3.0)	25 (2.4)	39 (1.5)	75 (0.8)	214 (0.3)	300+ (0.0)
OCT	286 (0.2)	94 (0.6)	47 (1.3)	33 (1.8)	29 (2.1)	31 (1.9)	39 (1.5)	67 (0.9)	158 (0.4)	300+ (0.1)	300+ (0.0)
NOV	300+ (0.1)	182 (0.3)	87 (0.7)	59 (1.0)	51 (1.2)	53 (1.1)	71 (0.9)	133 (0.5)	300+ (0.2)	300+ (0.0)	300+ (0.0)
DEC	300+ (0.0)	300+ (0.2)	143 (0.4)	85 (0.7)	67 (0.9)	70 (0.9)	91 (0.7)	162 (0.4)	300+ (0.1)	300+ (0.0)	300+ (0.0)

(Tables compiled from data supplied by the Air Resources Laboratory of the National Oceanic and Atmospheric Administration.)

SHADE AND MICROCLIMATE

To this point, we have seen the relationship between UV intensity and factors we cannot control—geographical, climatological, etc. If there is not much we can do to change the ecological factors with which we live, there *is* an area over which we have considerable control. This is our microclimate or microenvironment. If we find ourselves outdoors at noon on a sunny day in Quito, Ecuador, what are we to do? The first impulse might be to seek shade. Shade offers a microclimate, one quite different from the sun-drenched area outside it.

Because of the potency of skylight, we have to recognize that the protection from UVR offered by shade may be limited. Most people feel they are out of the sun when in full shade, but this is not so. As early as 1936, two employees at an Ohio sanitorium noted that patients in the full shade of the building got about as tanned as those in the sun. More recently, A. V. J. Challoner and his associates at the Institute of Dermatology in London conducted research that provides us with more precise information.[4] They developed a UV dosimeter, a piece of UV-sensitive film to be worn on clothing which gives an indication of one's level of UV exposure. They placed nineteen of these dosimeters around the body of a mannequin which was positioned, facing south, on a sunny lawn. At the end of the day, the film revealed that there was surprisingly little difference in the dose of UVB received by the front as compared to the back of the mannequin. So whether we sit in the shade of a tree or building, turn our backs to the sun, or sit beneath a beach umbrella, skylight may foil our attempts to limit UV exposure. Later in this chapter, specific recommendations will be made concerning how we may use shade to attain maximum protection when desired.

Reflection

An important determinant of the amount of UVR we receive is the type of surface we are on, be it grass, snow, sand, or water. Each of these surfaces reflects a portion of the UVR

striking it. The following table reveals the amount of total sunlight and the amount of UVB reflected from various surfaces:

	Reflectivity of Different Surfaces	
	% of Total Sunlight	% of UVB
Fresh snow	89	85
Old snow	50	50
Bright, dry dune sand	37	17
Bright, wet sand	24	9
Sandy grass area	17	2
Heather, berries	9	2
Water	9	5
Water above sand	12	10
White skin	35	1

Several points of interest emerge from this table. The proportion of total sunlight and of UVB that clean snow reflects
in considerable. The high percentage of total sunlight reflected helps explain why winter snow seems to stay on the
ground so long in spite of bright sunlight—little heat is being
absorbed. Rapid melting does not begin until the snow becomes dirty.

More important for our purposes, fresh snow also reflects
back 85 percent of the UVB, greatly increasing the amount
we are exposed to. The UVR that one receives standing on
snow is about double that received standing on bare ground.
Moreover, snow-reflected UVR strikes portions of the body
that are not usually exposed, such as the chin and the bottom
of the ears. These areas are particularly sensitive. This intense
UVR, in addition to the tremendous increase in visible light,
can cause snow blindness and permanent damage to the un-
shielded cornea of the eye. To protect against such harm, Eskimos from the Arctic have devised wooden slit goggles and
modern skiers wear sunglasses.

Water

A surprise for most people is the low reflectivity of water.
When the surface of water is smooth, it reflects but 5 percent

of the UVB, a slightly larger proportion if it is rough. When the sun is only 10 to 20 degrees from the horizon, however (about one hour before sunset), water reflects more than 50 percent of the UVB. But when the solar angle is this low, there is hardly any UVB to be reflected and little cause for concern.

Many are more worried about UV reflection from water than they need be. A recent article in an outdoor sports magazine advised its readers on the dangers of overexposure to UVR when fishing because, the article noted, of the high reflectivity of water. Some dermatologists warn their skin cancer patients about the dangers of water-reflected UVB. And the American Cancer Society cautions in a brochure on skin cancer that, when swimming, UV rays are "bouncing toward you from all directions—off sand, water, patio floor, deck." Why is this myth so prevalent?

One answer is that when it comes to UVR, what we see is not what we get. Some visible light is reflected from water, and at low solar angles the amount reflected is considerable and intense. But, as noted, there is little UV coming at us this late (or early) in the day. Another reason for the prevalence of this mistaken notion has to do with our experience of suffering painful burns while around water. Because water cools us, we can endure the hot sun for longer periods of time. When we get hot, a quick dip refreshes us and our exposure may be extended. Without the water, we probably would have headed indoors. Also, even while immersed we are getting about as much UVR as if we were on land. If water reflects so little UVR, it must transmit most of it, and such is the case— it penetrates the water and strikes our skin. Again, we don't notice because we are spared the heat.

The low level of UVB reflected from bright dune sand— only 17 percent—might surprise some people. The reflection from the partially vegetated sand of the typical desert would be considerably less.

Check again the reflectivity of white skin, only 1 percent. The rest is absorbed, mainly by the epidermis.

OTHER FACTORS: HEAT, HUMIDITY, AND WIND

Donald Owens and his colleagues at the Baylor College of Medicine have published a series of studies relating the effectiveness of UVR to various weather conditions.[5] High temperatures, they found, can aggravate any harmful effect of UV rays and can accelerate the process of tumor formation. Using mice as subjects, they found that those given UV irradiation after being heated 10 degrees Fahrenheit developed a more intense redness and crusting than did those at normal environmental temperatures. Mice in a cooled environment reacted even less than those kept in normal temperatures, and later tumor development was also delayed. This research confirms what has long been known for humans—that heat adversely affects photosensitivity reactions, causing more suffering for those who have allergic or toxic reactions to the sun.

Humidity also increases the damaging effects of UVR. In Owens's study, hairless mice (chosen because of their high sensitivity to UVR) were kept for ninety-six hours in a chamber with either 5 percent, 50 percent, or 80 percent relative humidity. They were then given either three or eight MEDs of UV irradiation from a sunlamp. Animals kept in high humidity chambers showed more damage after UV irradiation only when they received the ten-MED dose. Thus, humidity does seem to accentuate the damaging effects of UVR when exposure levels are high. In the natural environment, the UVR-scattering effect of humidity would tend to decrease the overall dose of harmful rays.

Owens and cohorts next took some mice and immersed them in water for six hours prior to UV irradiation. Now even at three MEDs, marked skin damage occurred. Other researchers have reported that with humans, immersion of the body before UV exposure decreases the amount of time it takes to get a sunburn. As the outer layer of the skin becomes thoroughly soaked, it allows the penetration of light that would normally be scattered or reflected. Wet clothing has the same effect. More UVR penetrates a wet T-shirt than a dry one, since water transmits radiation (which would normally be reflected) through the shirt.

Since wet skin burns more readily than dry skin, swimming decreases the time it takes to get a sunburn. Activities that make a person sweat may do the same, though there is a natural sunscreen in some people's perspiration.

Wind has similarly been found to augment the effect of UVR. Again using hairless mice, it was found that animals kept in wind chambers providing a velocity of six miles per hour for four weeks and given three MEDs of UVR daily showed greater skin damage than mice exposed to UVR alone. Animals exposed to wind but not receiving any UVR showed no damage at all. These results lead to the speculation that since wind causes no ill effects on its own, there may be no such thing as windburn per se. What we refer to as windburn is probably sunburn that has been intensified by the wind.

In sum, then, Owens's research demonstrates that a person is more likely to suffer a sunburn and other skin damage on hot, humid, windy days than on days when weather is cool, dry, and calm.

Atmospheric Aerosols

The intensity of UVR reaching the earth also depends on aerosols in the atmosphere. Aerosols, remember, are small particles introduced into the atmosphere by natural processes or human activities. They include water vapor, dust, volcanic particles, haze, smog, and other pollutants. These aerosols absorb or scatter some of the UV rays, thus decreasing the amount reaching the earth, and the difference between UV intensity in city and country can amount to 5 percent or more. Ultraviolet monitors placed in Philadelphia and Honey Brook, Pennsylvania (seven miles apart), revealed that the small town received 5 percent more UVR on an annual basis.[6] Philadelphia's air pollution is probably responsible for the decreased reading.

Astronomers at the Mount Wilson Observatory in California have been keeping records of the average intensity of sunlight and UVR for the past sixty years. They have noted a 10 percent decrease in the intensity of sunlight along with a 26 percent decrease in UVR due, they believe, to increasing pollution in their vicinity.

These observations stand in sharp contrast to the speculations on increasing UV levels. Recall the concern some people have had with the effects of SSTs, Freons, and pollution on the ozone layer (discussed in Chapter Two). Theoretically, these could drastically increase the amount of UVR allowed through the atmosphere. Actual observations, on the other hand, indicate relatively constant levels of UVR over the long haul and even some areas where pollution has caused a decrease in UVR.

HOW TO KNOW WHEN YOU'VE HAD ENOUGH

It takes many hours before a real sunburn appears on exposed skin. How are we to know, while we are in the sun, whether we have already had too much? The factors we have covered previously (season, time of day, etc.) should always be borne in mind, and it is very important that we be aware of our skin type. In addition, two simple tests can be made to get a rough indication of how much UVR the skin has absorbed.

Skin Test I: The first time we go out in the sun during a given year, we should be careful to note any difference in color between protected and unprotected areas of the skin. If the skin beneath a bathing suit strap or under the waistband of trunks is a different color from exposed skin, it is time to seek protection. Some physicians are unaware of this because research they have read generally involves producing erythema with UVB lamps. It takes hours for redness to develop with UVB. But in sunlight, the longer wavelengths are present also, and these produce immediate tanning or reddening. So, as one receives a considerable dose of UVA and visible light, one may at the same time be receiving too much of the more damaging UVB. The immediate reddening is thus an indicator of a potential later burn. Using immediate reddening as such an indicator is valid, however, only if there is considerable UVB *and* longer wavelengths in the daylight spectrum.

In the early morning and during the evening, one may receive enough of the longer-wavelength radiation to develop immediate reddening but at the same time receive only a small and harmless dose of UVB. On cloudy days, on the other hand, the amount of visible light and UVA is affected much more than is UVB, most of which reaches the earth in spite of cloud cover. Thus, the immediate reddening may be delayed by cloudiness while the real sunburn progresses at about the normal rate. We may get too much sun, in this instance, before noticing the color change that tells us to get out of it. This test, then, is best used when the sun is high and under mostly clear skies.

Skin Test II: If we have been in the sun for some time and want to know how much of the color we see in our skin is attributable to real burning or tanning and how much to immediate reddening or pigment darkening, another simple test can be performed. If pressure is applied to the skin and then relieved, skin that is truly burned will turn white, since the blood in the peripheral vessels is temporarily displaced by the pressure. If the color we see is due only to immediate pigment darkening, there will be no blanching of the skin when pressure is applied.

You might try this test the next time you are in the sun and have applied a high-SPF sunscreen product (15 or above). Most sunscreens only filter out the UVB (though increasingly they filter out UVA as well), but visible light still reaches the skin and causes an immediate darkening reaction in most of us. By applying pressure to the skin with the tip of a finger, you can assure yourself that this darkening is not the forerunner of a sunburn—no blanching will occur and the color will soon fade.

Dosimeters

These two techniques—observing differences in exposed and unexposed areas and seeing if the skin whitens with pressure—are less than perfect and in some cases only tell us when we have had too much sun after the damage is done. In recent years, polysulphone plastics, which darken when exposed to UVR, have been developed as a more sensitive measure. These badges, call dosimeters, can be worn on the clothing and give us very accurate information about the UVR we are exposed to.

A. Zweig and W. A. Henderson, Jr., of the American Cyanamid Company, have developed a practical dosimeter which is sensitive to radiation between 267 and 330 nm.[7] When UVR is absorbed, the film turns a reddish-orange hue and this color can be compared to a printed standard. The darker the film, the more UVR has been absorbed.

Dosimeters have been used with groups that typically get too little sunlight (such as the elderly, submariners, and people with dark skin who live in northern latitudes). Individuals in these groups have a higher risk of developing health problems associated with insufficient sunlight. With increasing interest in the potentially harmful effects of sunlight, we will begin to see a greater use of dosimeters among people who tend to get too much sun.

FIELD STUDIES USING PERSONAL DOSIMETERS. Various activities expose us to differing amounts of UVR. In a study undertaken in Australia by C. D. J. Holman and his associates, volunteers wore polysulphone dosimeters on five body sites for several consecutive days.[8] As expected, a classroom teacher received the lowest level of UVR—about 10 percent of that available—and a gardener and bricklayer the highest amount, up to 80 percent.

When different recreational pursuits of the subjects were considered, the order of activities from highest to lowest UVR received was as follows: swimming in the ocean, boating, sunbathing, hiking, golf, fishing, tennis, pool swimming, cricket, and gardening.

It is surprising that sunbathing does not head the list. Most sunbathers, however, take the rays in a prone position, so only half the body receives UVR at a given time. Any physical activity exposes one side of the body to sunlight plus skylight UVR, the other side to skylight UVR, thus increasing the total dose. But we must remember that dosimeters, not skin, were employed in this study. Skin does not burn as quickly when a person is active. A moving target absorbs only one-third of the UVR compared to a stationary one, and in fact sunbathing leads to greater effective doses than any activity.

The low ranking of pool swimming is accounted for by the low dose received by the legs during this activity. The shading by pool walls and the special properties of chlorinated water account for this. At a depth of two feet, only half the UVR is received.

The areas of the body, averaged over all subjects, that received the most UVR were the hand and upper back. The lowest exposures were recorded on the cheek. It is wise to keep this in mind when applying sunscreen.

Lifetime Sunlight Exposure

Some of the effects of sunlight are cumulative, and it is important to know how much exposure one has accumulated over the period of a lifetime. Occupation is one factor to take into consideration. While farmers, sailors, outdoor construction workers, and police officers are in a favorable position to reap the benefits of sunlight, they also have a higher incidence of skin cancer than office workers, houseworkers, indoor manufacturers, and clerks.

Lifetime exposure is related not only to occupation, but also to recreation, hobbies, and other factors. Frederick Urbach and his associates at the Temple University School of Medicine use the following criteria to measure lifetime sunlight exposure:[9]

1. time spent outdoors in nonoccupational pursuits such as sunbathing, working in the garden, hiking, or relaxing
2. outdoor activities connected with military service
3. job history, including time working outdoors
4. watching or participating in outdoor sports in the daylight.

Each of these factors is corrected for season and latitude. Urbach and associates found a relationship between total time in the sun and skin cancer, though not a perfect one—fully half of the nonmelanoma skin cancer patients had only minimal sunlight exposure times.

Now that we are aware of the factors that determine the quality of sunlight we find ourselves in at a given time and place, what are we to do about it? We can determine approximately how long it will take us to burn; we can take into consideration factors such as reflective surfaces, heat, humidity, and wind; we can use dosimeters to determine the precise amount of UV exposure over short periods of time and even calculate our total exposure through our lifetimes.

Obviously, we want to get enough and avoid too much sun. In wintertime and in northern latitudes, we are well advised to take walks in the sunshine when possible. When this is impossible or uncomfortable, just sitting in front of a sunny window allows the skin to absorb UVA and the whole body to receive the benefits of visible light. One may turn to artificial sources of UVR such as sunlamps and tanning salons, though the possible dangers associated with these techniques will be discussed in Chapter Six.

But what if your problem is a different one? You love the sun, want to seek its benefits, and enjoy the positive mood it instills. Or you love to hike, hunt, swim, ski, or sunbathe. We are known as a nation of sun worshippers, and for this reason there will be occasions when many of us will want some sort of protection while remaining outdoors in the sun for long periods of time.

Or you may suspect that you have already had too much sun through the years. If skin damage has occurred, there is all the more reason to begin protecting yourself. It is never too late to start a program of moderation. Given a chance, skin can regenerate and repair itself to a degree, and tumors that are in the premalignant stage (for example, the solar keratoses) will often go away if overexposure is discontinued. How, then, can we protect ourselves against too much sun?

PROTECTION

We should be thankful for the physical properties of plain window glass. Through it passes all visible light, and we can observe the sunlit world as we look through it. But glass blocks 95 percent of the UVB. Thus, the chemical (or actinic) effect of sunlight passing through glass is greatly diminished.

A. Roffo was the first to discover that he could not produce skin cancer in mice when sunlight was passed through glass. He had created tumors using unfiltered sunlight, but when ordinary window glass was placed between sun and animals, no cancers appeared. We need not worry, then, about time spent in our cars (if the windows are rolled up) or on our sunporches (if they are glass enclosed).

But frequently we want to get out in the sunshine. Clothing should be our first consideration.

Clothing for the Sun

When it comes to suiting up for the sun, several recommendations can be made. When the sun is high, between ten A.M. and three P.M. Standard Time, the person desiring sun protection should emulate those who have lived with a surfeit of sun throughout their lives—people like Arabs and working cowboys. Wearing long, loose-fitting clothing not only shields one from the sun's UV rays but also from the long, hot infrared rays. If clothing is loose, air is free to circulate beneath it and water can evaporate from the skin, thereby cooling it.

Clothing, of course, can be made from many different materials and also varies in color, structure, and thickness. Together, these factors determine the amount of protection from UVR offered by a particular fabric. In general, the fabric's structure outweighs all other factors. A tightly woven fabric—whether nylon, polyester, cotton, or wool—offers maximum protection. Woven nylon (tricel), for example, transmits less than 1 percent of the UVB whereas knitted nylon (acetate) transmits about 25 percent. Since most cotton fabrics are tightly woven, it matters little whether they are printed, denim, or needlecord. In each case as little as .1 percent of UVB gets through, depending on color.[10]

While color is of some importance, it is only a minor factor when compared to structure. A light-colored printed cotton, for example, transmits about 3 percent of the UVB compared to the .1 percent found with a dark-colored printed cotton. In either case the amount is small, and color should be chosen in accordance with temperature and not UVB.

A wide-brimmed hat is recommended, and it too protects the head from both UV and infrared radiation. The wider the brim the better. The cowboy hat can be made more effective by training the brim down—a curled-up brim allows too much sun on the face. A Mexican sombrero, again turned down, offers the ultimate protection, but gardening hats, panama hats, and the conical hats worn by Asian peasants are also good. Persons who are ultrasensitive to the sun may have to wear gloves as well when outside at midday.

Remember that clothing, when wet, transmits more UVR than it normally would. Of course, wetting the clothes cools us through evaporation, so one must decide whether heat or UVR presents the greatest problem at the moment.

Shade

Nothing can beat slipping into the shade on a hot, sunny day. Air temperature falls, there is some protection from excessive UVR, and one still enjoys all the benefits of being outside in the beauty and fresh air. But the coolness of shade, as you will recall, is no indicator of the amount of sunburning radiation coming our way. If the shade is heavy, the portion of sunlight coming directly from the sun is blocked, but the skylight and reflected light remain. Unless we are on snow or bright sand, reflected UVR does not amount to much. But the UVR in skylight is considerable and one may be receiving 50 percent or more of the UV rays that would be received sitting in the sun. And since we don't get as hot in the shade, we may be outdoors for a much longer time.

Sitting in the shade of a building, one receives skylight from over half the sky, the portion not blocked by the structure.

Sitting in the shade of a palm or any other small-crowned tree allows nearly all the skylight through, since only a tiny portion of the sky is blocked.

To get the maximum benefit from shade, it should not only be heavy but extensive in area as well. Sitting in a grove of trees, for example, can give almost total protection. If the crowns are spreading and touch one another, a perfect canopy is created and both direct sunlight and skylight are reflected or absorbed.

The outdoors person who is concerned about too much sunlight can seek the most complete shade available. On the desert or open plain, however, where the need for protection is great, the available shade may not be substantial. If you are concerned about excessive sunlight and find yourself in an area where natural shade is limited, it is wise to carry a light piece of opaque material that can be tied to plants and which makes a roof over your head for the hottest parts of the day. The larger the area of shade created, the better the protection will be.

As long as we understand what shade can and cannot do, and the differences in the types of shade, we can gain considerable comfort and protection by seeking it out when the sun is too intense. The only caveat is to watch out for the false security we may feel when thin shade gives relief from the heat—the sunburning potential may still be high.

Sunscreens

Sunscreens, when applied to the skin, absorb or scatter varying amounts of UVR. Sunblocks are opaque and reflect all UVR and visible light. Most people are now aware that the suntan lotions we used in years past gave little or no protection from UVR, but merely kept the skin moist while the sun did its work. As recently as 1975, the commercially available sunscreens offered only marginal protection. True sunscreens are now available and have recently come into wide use.

Among primitive peoples living in the tropics, in deserts, or at high altitudes, a number of sunscreens have been and still are used. Ochre and burnt-umber pigments, for example, are mixed in a cream base and spread over the skin, giving protection from both UVA and UVB. In some places, a coat of mud is applied as a sunblock.

Today we are aware of numerous chemicals that are effective as sunblocks and sunscreens. Zinc oxide is a white, opaque substance which blocks all light from the skin. Being opaque, however, zinc oxide is highly visible on the skin and cosmetically unappealing for many people. Sunscreens are invisible and therefore more acceptable.

Vitamin E and beta-carotene absorb UV rays and thus qualify as sunscreens, but the most effective sunscreen, and the one found in most commercially available products, is PABA (para-aminobenzoic acid) and its derivatives.

In 1975 it was first shown that a 5 percent PABA solution protected hairless mice from skin cancer and elastosis of the dermis. Products with PABA screen out wavelengths between 290 and 320 nm, the UVB band. Increasingly, sunscreens also include benzophenone and its derivative, oxybenzone, to extend the protection into the UVA range (peak effectiveness between 340 and 360 nm).

Sunscreen preparations can absorb or scatter up to 99 percent of the UVR, depending on the concentration of PABA and other active ingredients. The degree of protection is printed on the label as the Sun Protection Factor, or SPF. Safe exposure time can be determined by multiplying the SPF number by the number of minutes of unprotected exposure you can take before there are undesirable effects. An SPF of 8 allows one to stay in the sun eight times longer than if unprotected; an SPF of 15, fifteen times longer.

In studies with mice, sunscreens with an SPF of 15 or higher completely protected the connective tissue of the skin from change even under strong UV exposure. Thus, the onset of wrinkling may be delayed and we can improve our chances of avoiding skin cancer by using high-SPF products. There is even some evidence that by using sunscreens every day, a protective layer is built over the skin and previous damage may be given a chance to repair.

Sunscreen products are also available for lip protection. Lip balms with screening agents should be applied to this sensitive tissue.

Sunscreens for the '80s

Now that pharmaceutical companies have successfully developed high-SPF sunscreens that are cosmetically appealing, public demand for them is increasing rapidly. And researchers are continually developing new formulations with more protective properties. In 1978 Drs. Kays Kaidbey and Albert Kligman of the University of Pennsylvania Department of Dermatology evaluated the available sunscreens and found very few products with the high SPFs we can buy today. Of all the products tested in 1978, only three had SPFs greater than 10. They repeated their testing procedure in 1981 and found several formulations with, as they express it, "remarkable qualities."[11] Products with SPFs double that available only a few years ago were commonplace. Moreover, these new products provided protection for longer periods and resisted washing off with swimming or sweating.

But the SPF numbers on a product are often no more reliable than the MPG estimates found in automobile ads. Sunscreen manufacturers are required to demonstrate the effectiveness of their product, since sunscreens are considered drugs by the FDA because they are designed to protect the skin against solar damage. But tests are conducted under laboratory conditions and field tests have shown that many sunscreens have SPFs several numbers lower when used in natural conditions. No sunscreen currently on the market has an SPF greater than 12 when tested under field conditions.

Water in particular can reduce sunscreen effectiveness by as much as 90 percent and many manufacturers recommend a reapplication after swimming. Sweating similarly washes many sunscreens from the skin, but a natural component of sweat, urocanic acid, partially compensates for this. Urocanic acid is a natural sunscreen. No reliable SPF value can be given for it, however, since people differ widely in the amount of urocanic acid present in their perspiration.

Products that contain octyldimethyl PABA lose very little

of their effectiveness after more than five hours on the skin. Moreover, this type of sunscreen does not wash off after forty minutes in a Jacuzzi whirlpool. In contrast, products that contain PABA alone, especially if they have an alcohol base, wear off much more quickly and offer absolutely no protection after only ten minutes in a whirlpool. Unlike octyldimethyl PABA, PABA is absorbed rapidly into the dermis, leaving little to protect the epidermis. That which does remain is washed off easily, since it is water soluble. By contrast, octyldimethyl PABA is only poorly soluble and therefore stays on the skin rather than dissolving in water and being washed away.

Of the products tested by Kaidbey and Kligman, those containing octyldimethyl PABA together with oxybenzone offered the highest SPF, with ratings greater than 22. If you are looking for a high-SPF sunscreen, products with these two ingredients are a safe bet, but only when they are mixed in a cream base or a milky lotion. Clear lotions, typically containing alcohol, offer inferior protection at the time of application and also have poor resistance to wear-off and wash-off.

Theoretically, any product with an SPF of 15 or greater offers full-day protection from UV erythema in the middle of the summer at almost any latitude. But because of other factors mentioned (wear-off, wash-off, rubbing, etc.), some of the less effective products will have to be reapplied. And even those most effective against wash-off guarantee protection for only forty minutes of swimming. With prolonged swimming (and even sweating), reapplication is necessary.

In the 1980s, we have sunscreens that offer as much protection from the sun as does our clothing. The terry cloth material used for making beach robes, for example, has an SPF of about 15. With proper use of sunscreens, we are nearly as protected as when wrapped in our robes.

Choosing Your Sunscreen

The following table lists the SPF you should look for in a sunscreen for your type of skin.

SKIN TYPE AND SPF REQUIREMENTS

Skin Type	SPF Required
I	10–15
II	6–12
III	4–6
IV	2–4
V	——
VI	——

Problems with Sunscreens

There can be little doubt that sunscreens are effective in keeping potentially dangerous radiation from the skin. But remember that our sun is a paradoxical star which can confer both benefit and harm. Unfortunately, the very wavelengths filtered by sunscreens are those that kill bacteria and stimulate vitamin D synthesis. We lose these benefits as we gain protection. For a while some European sunscreen manufacturers tried to remedy this problem by adding vitamin D to their products. This didn't help much, since the body cannot utilize vitamin D administered in this manner.

If we apply suncreen *after* we have allowed a modest amount of sunlight to fall on unprotected skin, we can attain the benefits and avoid the dangers.

There are some other problems with the chemicals in sunscreens. PABA can cause a drying of the skin after prolonged use, but the use of moisturizing creams after the sunscreen has been removed helps, as does applying these creams before the sunscreen.

Sunscreens brought to the eyes in sweat is an especially irritating combination. For those who perspire heavily, a solution is to apply sunscreen to the lower part of the face only, relying on a hat to protect the forehead and eye area.

On sensitive people, PABA and its derivatives can cause a stinging and flushing of the skin. Products such as Sol Bar Plus and For Faces Only are available for those whose skin reacts to the standard PABA preparations. In addition, some of the problems can be avoided if sunscreen is applied indoors while the skin is cool and dry. Sweaty skin also makes an even application difficult. And by shaving the night before an outing or skipping a shave altogether, the skin will be much less sensitive to PABA irritation.

It is also a good idea to let a sunscreen product dry completely before putting clothes on over it, since PABA can cause a yellow staining (usually washable). Later sweating, however, may deposit sunscreeen on the clothes anyway, so the wearing of older clothes when possible is recommended.

Future Sunscreens

Pharmaceutical companies are working on the development of oral medications which are photoprotective. Nearly all of the objections to today's sunscreens would be answered except for the fact that the benefits of UVR would still be lost. The psoralens may someday offer protection in the sunburn range (290–320 nm), and beta-carotene, taken orally, may give protection in the visible range (400–720 nm) for the minority who are especially sensitive to these wavelengths.

THE PROTECTIVE TAN

Many dermatologists think that the words "healthy tan" are a contradiction. For them, a tan (or burn) indicates overexposure and therefore damage to the skin. But once a tan has been achieved, it can offer some protection against future sun-induced damage. (This is not to claim that the darker you get, the safer you are. The damage done to the skin in developing and maintaining a dark tan, especially for those with the more sensitive skin types, is considerable. So while a tan can protect against future overexposure, it does indicate that some damage has already been done.)

A tan protects against future sun damage in much the same way that naturally dark skin does. Black skin takes on its color and gets its characteristic protection from sunlight because of the number and pattern of melanosomes (the granules that contain melanin) in the skin. Black skin can have four times the number of melanosomes as untanned Caucasian skin, and in blacks these melanosomes are distributed more evenly throughout the skin than is the case with whites. Thus, blacks have more protective melanin and a more even distribution of it.

Ultraviolet-B causes the production of melanin in the skin and also creates a more even distribution of the melanin, giving a more continuous coverage. Melanization, then, is a natural form of photoprotection. In addition to melanization, UVB also thickens the horny outer layer of the skin, the stratum corneum, so that less radiation passes through to the deep layers. (Parenthetically, though scientists once believed that black skin had a thicker stratum corneum, this is now known to be untrue.)

So a tan can protect against a future sunburn. Does it also help us to avoid skin cancer and wrinkling? For individuals who are sensitive to the sun, probably not, though we will look at a qualification of this statement later in the chapter. Unless extreme caution is taken, the amount of UVB absorbed in achieving and maintaining a dark tan is more harmful than beneficial. This does not mean that these people should avoid the sun altogether, of course, but just exercise some of the precautions discussed earlier in this chapter. For persons who have demonstrated sensitivity by developing a skin cancer, intentional sunbathing should be discontinued and protective measures taken. The symptoms—the skin cancers— are sending a message that too much sun has already been received. Such people are best advised to forgo the "healthy tan."

A Sensible Sunning Schedule

Evidence is accumulating that all of us, no matter what our skin type, would do well to understand the effects of sun on our skin. This does not mean that we should become so paranoid that we curtail our outdoors activities in favor of the ultimate photoprotection, darkness. But being aware of the relationship between time of day and UVR, the importance of proper clothing, and the utility of sunscreens might well prevent later grief.

When one examines the evidence, it becomes increasingly clear that there is little justification for sunbathing. Most of us can get all the sunlight we need by simply planning some outdoor activities—whether a walk, jogging, biking, or gardening—when weather permits. The tonic effect of *activity* in sunlight cannot be overestimated. And one can stay in the sun much longer when moving than when stationary—about three times longer—before there is damage.

But for those who insist on sunbathing, some recommendations can be offered. First, in the summertime it is best to avoid sunbathing between the hours of ten A.M. and three P.M., Daylight Savings Time, when the sun is most intense. Second, if sunbathing is done during the high-sun periods, the first day's exposure should be limited to fifteen minutes with five minutes added each day until a tan is acquired. (The first day's exposure could be for as long as forty-five minutes if you are moving around.) Third, don't forget that time in the water also counts as exposure time, since the water transmits most of the UVR, letting it hit your submerged body. Fourth, after the maximum exposure periods have been reached, apply sunscreen if you wish to stay in the sun and do not want to put long, protective clothing on. With clothes on, sunscreen on the face is a good idea as well as on the V of the neck and upper chest if an open-collared top is worn.

With the use of a little sense in the sun, we can gain the many benefits of sunshine while avoiding its dangers.

The Suberythemal Dose

Many people wonder whether they can attain the benefits of sunlight and avoid the problems by limiting their exposure to a period of time just under what it takes to receive one MED or, as photobiologists say, a suberythemal dose. As we have previously seen, the minimum erythemal dose (MED) is the length of exposure to a certain quality of sunlight that produces a barely perceptible redness in the skin. This redness is noted at the peak of sunburn intensity, usually twenty-four hours after exposure.

A suberythemal dose would cause no problems for most

people. It is when we get five or ten MEDs that the real trouble begins and we suffer the painful consequences. And it is doses of this magnitude that, when repeated too frequently, can set the stage for the more serious long-term consequences.

But recent evidence suggests that even with suberythemal exposure, some epidermal cells die or suffer damage to the genetic material, DNA. On the positive side, such low doses are capable of stimulating the development of melanin, so one may acquire a tan over several days without ever seeing a hint of redness.

One cannot fool Mother Nature by exposing oneself to several suberythemal doses in the course of a day. By going out in the July sun for fifteen minutes at eleven A.M., one P.M., and three P.M., and then perhaps repeating this schedule the next day, you will receive a cumulative dose far in excess of one MED. Any sunlight received lowers the amount of time it takes to burn during any subsequent exposure within the next twenty-four hours.

So You Still Want a Tan

Within the next decade, I believe we will witness a return to standards of skin beauty that will approximate those of the last century. During the 1960s and 1970s many of us got as dark as we could during the the summertime, and increasing numbers extended the tanning period to twelve months a year by using sunlamps and, more recently, tanning salons. Tanned skin, we believed, looked good. But the profligacy of this behavior is becoming evident as we begin to notice the prevalence of the uncosmetic results—wrinkles and skin cancer. As the connection between these consequences and overexposure to sunlight is more firmly made through our experience, it is likely that the dark tan will be looked upon with about as much enthusiasm as heavy cigarette smoking.

We have been well aware of the effects of smoking for years, and most people now view the smoker with concern or pity rather than with admiration. And as the consequences of overexposure to sunlight—especially for fair-skinned people—become better known, the tanning addict may elicit similar emotions in us.

In the mid-1980s, however, many still insist on a tan. Tanning contests are held in some parts of the country with a prize awarded to the person exhibiting the most solar-induced damage. Even people of more moderate aspirations complain about how pasty they look in the winter and are not about to forgo the opportunity to remedy this come summer.

We can adopt a sensible tanning schedule and thereby minimize the inherent hazards. Exposing oneself to less than one MED of sunlight on a given day *might* have some harmful cumulative effects over time. But the verdict is not yet in on this, and for now we can say that it offers the least dangerous strategy for achieving a tan if a tan is desired.

John Parrish and his associates at the Harvard Medical School have recently published some research that bears on this issue.[12] While by no means dismissing the possible dangers, they do show that a person can tan, albeit slowly, without incurring the well-known damage of erythema.

We know that while UVB is responsible for both sunburning and tanning, it is more efficient in producing a burn. And while UVA is much weaker in this respect than UVB, it too promotes both sunburning and tanning, though it more efficiently induces tanning.

When Parrish et al. exposed subjects to UVB doses in excess of 70 percent of their MED for several days in a row, they began to develop sunburns. While the dose on any given day was not sufficient in itself to promote erythema, the cumulative doses were. Thus, people wishing to get some sun each day are advised to keep the dose below 70 percent of their MED if they wish to avoid sunburn.

More interestingly, when subjects were exposed to subthreshold doses of UVA (doses lower than those necessary to promote melanin synthesis and a tan), they began developing tans after a few days. (The tanning process, remember, is very slow and was determined in this study a full week after exposure.)

So subthreshold doses of both UVA and UVB eventually

result in tanning and burning, respectively. But the threshold for UVA decreased much faster—three times the rate of the threshold for UVB. That is, the length of time it took to induce tanning was dramatically decreased with each successive day's subthreshold UVA dose. The length of time it took to induce burning with UVB was also decreased, but to a smaller degree.

Though this research employed artificial light sources, we can easily see the implications for those who would like to apply this strategy out of doors in the sun. Ultraviolet-A does not vary as much throughout the day as does UVB. In the late afternoon—after three o'clock—UVB begins to diminish rapidly with each passing hour. Ultraviolet-A, since it is not scattered as much by the atmosphere as UVB, does not decrease proportionately and its intensity persists even as the day grows late. Sitting in the late afternoon sun for a period of time insufficient to promote a tan on a given day, we can start the tanning process going after a few days. And exposure times can be briefer and briefer, since the threshold for tanning decreases daily. The actual time spent in the sun would be dictated by MED—that is, how long it would take you to burn at that hour. A check back to the charts developed from the SUV meter will give you the number of minutes before burning for average skin, and Parrish's study suggests that you keep your exposure to less than 70 percent of this figure to avoid a cumulative sunburn. By using a UVB sunscreen, exposure time could be prolonged.

Drs. S. W. Becker and V. Vaile, dermatologists in Winter Haven, Florida, recommend a similar procedure, along with the use of a photosensitizing drug, for their patients with sun-sensitive skin.[13] Though their proposal is at variance with mainstream dermatological thought on this subject, it is worth taking a look at.

They begin their argument with the well-known observation that blacks have much lower rates of skin cancer than whites. As noted previously, black skin is characterized by the presence of many large melanin granules scattered evenly throughout the epidermis. In addition, Becker and Vaile subscribe to the theory that skin cancer involves a deficiency of the immune system such that the body does not eliminate malignant cells as they arise. (This viewpoint was discussed in Chapter Four.) Tyrosinase, produced by the same cells that synthesize melanin, is suggested as the specific antidote that protects the skin from carcinogenic damage from UVR.

Becker and Vaile give their sun-sensitive patients one of the psoralen drugs and have them sit in the late afternoon sun. Exposure time is limited to prevent sunburn. The psoralen is a photosensitizer, and before long a tan develops. The melanin granules produced with the aid of psoralens are larger than normal and tend to be evenly distributed in the skin. Moreover, the tyrosinase system is activated. The use of psoralens once a week maintains the effect.

The dermatologists report that they have been using psoralen-sunlight therapy on their vitiligo patients for twenty-five years and have noted no increase of skin cancer or other problems. They do not, however, report the consequences of using this technique with patients who have already developed skin cancer. This evidence remains to be collected.

Psoralens are prescription drugs, so psoralen-sunlight tanning is available only to those who can convince their doctors that they need it, a difficult task for anyone without obvious photosensitivity. But the slow, careful exposure to subthreshold doses of UVA is a procedure that anyone can practice. It is recommended for those who want a tan but wish to minimize sunlight-induced damage to the skin.

6

THE CIVILIZED SUN

Why, so this gallant will command the sun.
—William Shakespeare, *The Taming of the Shrew*

SINCE the invention of the incandescent lamp in 1879, our activities have become increasingly independent of sunlight. Artificial light, more than any other discovery, was responsible for the rise in productivity we have experienced in the modern world. Work is no longer restricted to the daylight hours and machinery does not have to sit idly during the night. But at what price? While humans have recovered a portion of the night to do with as they please, the quality of light we receive from artificial sources is very different from that to which we adapted over the millennia.

What Hath Edison Wrought?

The incandescent bulb was developed for one purpose—to allow us to see in its glow. But while vision is certainly important, light has numerous other influences on the body, from enhancing sexual functioning to improving liver activity. The wavelengths responsible for these nonvisual phenomena include the UV range and frequently extend through the entire visible spectrum as well. Ninety percent of the radiant power of an incandescent bulb, however, is in the infrared range, making it more accurate to call the light bulb a heat bulb.

Fluorescent lights emit in a similarly restricted spectrum. By the time they were developed, it was known that the eye's maximum sensitivity lies at a wavelength of 555 nm, and this is where fluorescent lights concentrate their output. They are truly vision tubes.

Artificial lights, then, differ greatly in spectral composition from natural sunlight, the light under which our ancestors evolved their daily rhythms and other light-related physiological functions. But natural and artificial light differ in other ways besides wavelength.

The intensities of sunlight and artificial light are very different. On a bright summer's day, sunlight intensity may reach 100,000 lux. The light in an indoor work area, on the other hand, rarely exceeds 2,000 to 3,000 lux, and if it did, the brightness would be most uncomfortable. So while we may be in a lighted environment for longer periods of time than our ancestors (that is, we have extended our photoperiod), we are exposed to light of far less intensity. Illumination researchers have focused their attention on developing the brightest light at the lowest cost without asking whether light so different from sunlight might have negative consequences for our health.

The revolution in lighting has been a mixed blessing. I will discuss the changes it has caused in the growth and behavior of humans as well as some of the problems associated with it. But artificial light also has its benefits, as any poultry farmer who uses it to increase egg production during the winter knows. In addition, artificial lights can be used to treat certain ailments and may actually prevent disease when used prophylactically with people who live in latitudes with long, sunless winters.

CHANGES IN GROWTH AND BEHAVIOR

Richard Wurtman, a photobiologist at the Massachusetts Institute of Technology, compares the change in our light environment over the last one hundred years to the near-universal administration of a new drug.[1] He notes that while vast sums of money are spent to study the effects of new food and drugs which we consume, minuscule amounts have been spent to determine the health-related correlates of switching from sunlight to artificial light. During the three and a half billion years since the advent of living matter, sunlight has been uniformly administered to all organisms at a given latitude. Since the development of artificial lights, however, the spectral composition and daily dosage of this "drug" have changed dramatically.

In recent decades, a sharp acceleration in the rate of growth in children has been noted in all northern countries. Researchers such as Fritz Hollwich speculate that, in addition to improving nutrition, this growth trend may be due to the increase in photoperiod we experienced with the discovery of the light bulb.

An equally noteworthy trend is the tendency for people living in temperate latitudes to reach puberty at earlier ages. And this effect is more pronounced in areas with access to artificial lights. In Russia, for example, girls living in urban areas reach puberty earlier than those who live in the country under more natural lighting conditions. Similarly, rats raised in the laboratory attain sexual maturity earlier than their wild counterparts.

As noted, with the use of artificial lights we experience a longer photoperiod, and spectral composition of light bulbs and tubes differs from sunlight. Hollwich and an associate found that light with wavelengths in the longer end of the spectrum (where incandescent bulbs concentrate their energy) had profound effects on the sexual maturation of drakes.[2] Light in the orange region caused their testicles to enlarge to six times that of birds in natural light. Red light caused a sixteenfold increase in size. Obviously, no similar experiments have been done with humans, but the decreasing age at which young people reach puberty these days is consistent with the animal research.

Sexual *behavior* may also be altered by artificial light. In Chapter Three the effect of sunlight entering the eyes on testosterone, the hormone responsible for our sex drive, was noted. John Hartung, a Harvard anthropologist, believes that humans in the developed countries have experienced an increase in the average level of testosterone produced because of the increasing photoperiod furnished by artificial light.[3] Photoperiod does have an effect on the activity of the gonads of all high-latitude mammals, birds, and reptiles studied to date, so the hypothesis is reasonable.

Consider the newborn infant emerging from the darkness of the womb to the superilluminated environment of the hospital. The lights are not only intense but continuous, since the baby is soon moved to the around-the-clock illumination of the nursery. Research shows that newborns have highly active pineals capable of producing much melatonin, a gonadic inhibitor. This is to be expected, given the absence of light in the prenatal environment. Intense, continuous lighting would be expected to decrease pineal activity. William Masters and Virginia Johnson, the sex researchers, have observed that sexual response—erection with baby boys and vaginal lubrication with newborn girls—occurs within a short period of time following birth (minutes for boys and within a day for girls). Could this be due to the light's effect on the pineal? There is no direct evidence for this, and with adults, light intensities must approach the levels of sunlight to inhibit pineal functioning. But effective intensities may be lower for infants, and this remains an intriguing possibility.

PROBLEMS CAUSED BY ARTIFICIAL LIGHT

Accelerated growth and sexual maturation induced by artificial light are not problems per se. But the stress reactions associated with this lighting are.

Outdoors, the quality and brightness of light vary from the dim, red glow of sunrise or sunset to the maximum intensities of midday. Sunlight also changes as clouds pass between us and the sun or as we change the direction of our gaze. Our natural light environment is constantly changing. When we work or relax in a lighted indoor setting, however, the situation is very different. Now the intensity and spectrum of light are unvarying for long periods of time.

As we go into school or office, we enter what has been called a "light cage." Fluorescent tubes are installed in bands against a white ceiling, creating uniform, monotonous lighting throughout the space. The depth and shadows seen when in natural light are lost. In the winter, exposure to unnatural light may be extended throughout the waking hours.

The eyes, to work properly, require constant variation in the light source. Without change, the pupils do not respond normally and visual purple (broken down when light strikes the rods in our retina) cannot regenerate at its normal rate. Similar problems arise when illumination is either too bright or too dim, and eyestrain is the inevitable consequence.

Eyestrain

The symptoms of eyestrain are numerous and varied. They include an inflammation of eyes and lids as well as a hot, itchy feeling in the area around the eyes. The visual system also breaks down and vision may be blurry or double. But symptoms also radiate out from the visual system—eyestrain can cause headaches, dizziness, and a nauseous feeling as well.

When the task before us is insufficiently lighted, when the light source is too bright, or when reflected light causes glare, muscles become strained. The stress affects the muscles that control the size of the pupil as well as the ones that change the shape of the eye as it converges and accommodates for near and far vision, respectively. The strain is even greater for those with a visual defect such as nearsightedness, farsightedness, or astigmatism.

Other Problems

As we enter the work place or school in the morning, the bright, artificial light triggers the production of cortisol, which has a stimulating effect on the body. This serves as a fine pick-me-up in the early morning hours. As the day wears on, however, we pay the price of the earlier "light fix" as premature fatigue sets in. This is offset somewhat by the continuous, monotonous light that radiates our way. Such constant, artificial stimulation, however, may well result in insomnia at night, since we are deprived of any real relaxation during the day and on into the evening.

Hollwich has exposed subjects to fluorescent lights for two weeks and noted a steady rise in the level of ACTH and cortisol, both stress hormones. The high levels persisted for two weeks after subjects were returned to normal daylight. The "light pollution" of artificial lighting may be responsible for all manner of stress-related symptoms from high blood pressure to emotional disturbance. Gastrointestinal problems, aches, and pains, as well as poor work performance can be other side effects.

Hollwich urges us to utilize natural light whenever possible—to work and play outdoors. We should depend on natural light and use artificial light only as a supplement. Where artificial light is necessary, says Hollwich, the intensities should be reduced far below the overstimulating levels we are accustomed to and should be subject to variation by use of a stage-selector switch.

Several other researchers are calling for increased study of artificial light's effects so that we may learn of potentially harmful biological consequences. One such project currently under way is exploring the relationship between fluorescent lights and melanoma.

MELANOMA AND ARTIFICIAL LIGHT

In Chapter Four it was noted that sunlight was *one* of the factors implicated in the causation of melanoma. Several studies have reported a correlation between geographical location and this type of skin cancer, and while melanomas can appear on parts of the body protected by clothing, they are more frequent on those areas that would be exposed when wearing swimwear. But some puzzling findings have come to light. Studies in both Britain and Australia have discovered higher rates of melanoma among professional people and office workers and lower rates among those working outside. Furthermore, while those outdoor workers who do develop melanomas tend to do so on sun-exposed portions of the body, office workers are more likely to develop them on the trunk.

Valerie Beral of the London School of Hygiene and Tropical Medicine, along with colleagues, reports that men and women who work under fluorescent lighting at their jobs have at least twice the expected incidence of melanoma.[4] Those who worked outside at any time in their lives actually have a lower than expected rate. The group experiencing the lowest risk of melanoma had neither worked outdoors nor in an office with fluorescent lights. But among the group who had spent some time in offices with these lights, the lowest risk was for those who had spent the most time outdoors.

Several points of interest emerge from this study. First, fluorescent lighting in the office may be an injurious form of illumination. (Television and fluorescent lighting in the home, by the way, were unrelated to melanoma.) In the UVB range, fluorescent lights produce peaks of radiation at 298, 302, and 313 nm. Tests reveal that many of these lights emit radiation at 290 nm at levels in excess of those recorded in Sydney, Australia. Being exposed to these wavelengths without receiving the rest of the UVB spectrum may be especially injurious. Certain wavelengths of light can repair the damage done by UVR, and though the specific ones that are effective in this manner are not known, it is quite likely that artificial light does not include them.

Second, since fluorescent lights do not initiate tanning—with the concomitant melanin production and the thickening of the skin—people exposed to fluorescent lights do not develop the protection from UVR that occurs in the sun. Perhaps this explains why office workers who had a history of sunlight exposure were not as likely to develop melanoma.

As long ago as 1940, a study appeared which suggested that exposure to sunlight—even overexposure—while tending to cause basal-cell carcinoma and squamous-cell carcinoma, resulted in a relative immunity to more serious forms of cancer. Little work has been done in this area since, and it warrants a reinvestigation.

Third, the finding that melanomas most frequently develop on the trunks of office workers reminds us that UVR penetrates our clothes. As much as 50 percent gets through the open-weave or light-weave clothing worn by women in the summer. The businessman's white shirt can transmit as much as 20 percent. Thus, the UVR in fluorescent light can reach the skin. Moreover, the trunk of the body is less likely to be exposed to sunlight, and if tanning does have a protective function, this area would consequently be the most vulnerable to fluorescent lights.

The research by Valerie Beral and associates will have to be replicated before we can make any confident conclusions about fluorescent lights and melanoma. Their study was actually designed to see if a relationship existed between melanoma and oral contraceptives—none did. The findings concerning fluorescent lights were serendipitous. Though their research is intriguing, a study specifically designed to test the light-melanoma hypothesis is still needed.

But if Beral's hypothesis is correct, how can it be reconciled with the evidence, presented in Chapter Four, of sunlight's role in causing melanoma? Beral found that, among those now working under fluorescent lights, having previously worked outdoors reduces a person's risk of developing melanoma. If a person had never worked under fluorescent lights, however, having worked outdoors is associated with a higher than normal risk of melanoma. Thus, in the absence of significant exposure to this type of artificial illumination, increased sunlight exposure is associated with higher rates of melanoma. Among those who had worked under fluorescent lights, their previous outdoor experience may have "hardened" them to the UVR of the lights.

While the effects of artificial light on the organism are not fully understood, research is under way to insure that we will get as much natural light as possible even when indoors, or when this is impossible, to make artificial lights that more closely approximate the light coming from the sun.

NATURAL AND FULL-SPECTRUM LIGHTING

At one time a critical task for architects was to design their buildings so that natural light could enter. Candles, kerosene lamps, and gas lamps were used only during the night, and illumination from these sources was much inferior to daylight. Buildings were usually narrow, since each room or office needed to face the outdoors. Ceilings were high and windows were large to let in all the sunlight possible.

With the development of artificial lights and especially fluorescent tubes, windowless offices, factories, and schools became possible. Buildings became boxy, and since windows were no longer necessary, much of the space was within the sunless interior. The inconstant light from the sun did not have to be relied upon as unchanging, shadowless illumination shone from the ceiling.

Today there is a revival in the architecture of the past. Energy has become a problem and designers are rediscovering the sun for lighting and heating modern buildings. And environmental psychologists are telling us that the most important factor in people's satisfaction with their work or school space is the presence of windows. The enclosed interior spaces of buildings are particularly annoying. As a result, we increasingly see these portions of buildings opening to sun-filled courtyards. And designs once thought to be archaic and extinct are reappearing. The clerestory, a vertical wall of windows adjoining the roof of a structure, is back in favor among architects.

In Las Vegas, New Mexico, the Department of Human Services' new building, with high ceilings and clerestories oriented to the east and west, was designed for maximum use of daylight without allowing overheating. The east-west orientation of the clerestories captures the early and late sun when light intensities are low. Since the building is used only in the daytime, the use of artificial lighting has been greatly reduced, and energy costs for the building are one-half what they would normally be.

But most buildings, no matter how well they are designed, must rely on artificial light to some extent. Moreover, light coming through the windows is deficient in the very active radiation lying in the UVB band.

As a result, some large plastics companies have begun manufacturing material for windows and skylights that differs significantly from ordinary window glass—it transmits wavelengths in the UVB range. Now, even though behind a window, a person can benefit from all the wavelengths present in natural light. (Concomitantly, of course, UVB-transmitting windows raise the same concerns that natural sunlight does.) Because eyeglasses and contact lenses similarly limit UVR, lenses made from materials of a similar type have been developed.

But until recently, the artificial lights still necessary to our modern society delivered radiation of a very different composition from that of the sun. The solution was the development of full-spectrum light (FSL), which more closely represents the wavelengths present in sunlight. These sunshine-simulating lights have attracted much research interest.

Hollwich, for instance, discovered that even when people are exposed to continuous high-intensity levels of FSL, the stress reactions that occur with a similar regimen of regular lighting do not appear. FSLs stimulated the production of less of the stress hormones ACTH and cortisol.

Other research has revealed that under FSLs in the factory, absenteeism and accident rates decrease while worker productivity increases. In a school setting, children achieve more and there are significant decreases in hyperactivity and other behavior problems.

While UVR is present in the FSLs, its intensity is low and no health problems have been noted. Soviet researchers have even reported an increased resistance to disease when a small amount of UVR is added to regular lights, as we saw earlier. But as we might expect, when UV intensity is increased to many times this level, problems occur. Such is the case with sunlamps used in the home and in the recently developed tanning salons.

SUNLAMPS

About one million sunlamps are sold each year in the United States. These lamps are rich in UVR and offer the user the opportunity to tan in less time than it takes in the sun—and to maintain that tan year-round. But thousands of injuries from these lamps are treated in hospital emergency rooms each year. Most commonly, they consist of acute sunburn and photokeratitis, or inflammation of the cornea. Some sunlamps provide UVR in the shorter wavelengths that are filtered out of natural sunlight by the ozone layer. Dermatologists are wary of the effects of these lamps and express great concern over the future skin cancers that may result.

Most of the problems with sunlamps arise from the indiscriminate use of them as tanees exceed the recommended time limits of exposure either intentionally or by falling asleep under the lamps. This was one of the motivations for the invention of the tanning salon, a business establishment where the person could receive controlled, supervised doses of tanning UVR.

TANNING SALONS

In 1978 tanning salon franchises first opened in the United States. Tanning salons started in Europe, where they were especially popular in northern Germany. During the long winters in these high latitudes, some ersatz sunshine brought a welcome respite. It is a little more difficult to explain why they developed where they did in the United States—in the South.

Tanning salons were first introduced in the Little Rock and Memphis areas and spread rapidly to other parts of the Sunbelt. Arizona has many, as does California. And surprisingly, southern California has proportionately more than northern California. The density of these salons in areas with ample sunshine is a puzzle for armchair sociologists to ponder—perhaps people in sunny places are more accustomed to being tanned and are reluctant to return to their pasty color for even a few months. Or, since people in the Sunbelt states can, with persistence, maintain a tan even during the winter, a pallid appearance may be more of a social liability in these areas. Those unable to spend most of their time in the sun can, nonetheless, sport a year-round tan by frequenting the tanning salons.

Whatever the reasons, there are approximately two thousand of these establishments operating in this country today.

While they may well be safer than the home use of sunlamps (and sunlamp-related injuries have decreased dramatically in recent years), tanning salons present their own hazards. Since the UVR inside a tanning booth may be ten times as strong as noontime-summer sunlight, dermatologists have raised their voices virtually in unison to condemn them.

Two types of tanning salon are currently in operation. One, the UVB salon, uses a fluorescent tube that emits radiation most strongly between 290 and 340 nm, with a peak at 313 nm. The second type, the UVA salon, uses tubes that emit in the longer UV range, the UVA.

When tanning booths came to this country in the late 1970s, the usual arrangement had the tanee standing in a booth with between eight and sixteen Westinghouse fluorescent lamps in it. Today the tanee often lies on a bed with as many as forty-eight lamps under it, sometimes with lamps above as well ("like a grilled cheese sandwich," comments the *FDA Consumer*).

UVB Salons

Many reports of eye damage (conjunctivitis, photokeratitis) have been reported by users of the UVB salons. The wearing of goggles designed to eliminate all UVR would prevent this, but this precaution is often not observed. Other hazards include reactions of people who are photosensitive either because they suffer from one of the light-sensitizing porphyrias (discussed in Chapter Four) or because they have taken or applied one of the many photosensitizing chemicals present in foods, medications, soaps, or cosmetics. The drinking, for example, of carrot juice, celery juice, or lime juice is photosensitizing for many people.

In the long run, damage may be more serious. Since UVB is associated with skin cancer, and since the "healthy tan" is really a reaction of the skin to UV damage, the year-round visits to the tanning salon become a way of maintaining cellular damage perpetually rather than giving the skin the break it usually gets in the winter months.

John Epstein, a San Francisco dermatologist, has pointed

out several false advertising claims made by the UVB salons.[5] Some of their brochures aver that the salons have eliminated the harmful rays present in sunlight and use only the beneficial mid-UV range. The shorter and longer wavelengths of the solar spectrum are not present, or so it is stated. The claim is wrong on two counts. First, the shorter UV wavelengths are not eliminated, since the radiation range of the tubes used extends down to 290 nm, as short as or shorter than any radiation naturally encountered on earth. Second, eliminating the longer wavelengths is not a health benefit unless the person has used a photosensitizing drug or has one of the porphyrias. Visible rays present no problem except in very rare cases.

The brochures sometimes claim that tanning is possible with this technique even if the person does not tan in the sun. This is not true, since UVB is UVB whether exposure is indoors or outdoors. If one cannot tan in the sun, one cannot tan under lights.

UVA Salons

In the UVA tanning establishment, introduced in Europe and gaining in popularity here, only 2 percent of the radiation is in the UVB range and the longer UVA predominates. This type of radiation is not nearly as dangerous as UVB, but neither is it as effective in tanning. UVA does have the advantage of promoting a tan without burning, but to initiate a lasting tan, about one thousand times more UVA than UVB is required. In the last chapter we saw that it is possible to acquire a tan, albeit slowly, by exposing oneself to *subthreshold* doses of sunlight UVA over a period of several days. But in the tanning salons, doses are decidedly *superthreshold*. Such massive doses can produce cataracts, and eye protection is therefore needed.

The most likely result of consistent use of the tanning salons of either type is the development of damaged skin, leading to premature wrinkling and, more rarely, skin cancer.

Peter Hersey and his associates have tested solarium users after twelve half-hour exposures on consecutive days.[6] They noted several changes in the skin's immune system, changes that appeared after exposure and which persisted for two weeks. In particular, there was a decrease in the cells that help guard against skin cancer.

CHEMICAL TANNING

There are ways of tanning that involve neither sunlight nor artificial light. Sunless tanning products (such as QT, Sudden Tan, and Indoor/Outdoor) contain dioxyacetone which forms a brown complex with proteins in the skin. Melanin is not involved and so, while people look darker, they need to take the same precautions in the sun as they would on their first day out.

A newer development is the tanning pill. Just as drugs were introduced a few decades ago that cured some of the ailments formerly treated with sunlight, a drug can now simulate what many people consider the only remaining benefit of sun-shine—tanning. Whereas sunless tanning lotions sometimes produce disappointing results—leaving the user with streaks and blotches—the tanning pills produce uniform coloration.

Two food colors are responsible for the chemical tan. Beta-carotene—which most of us consume every day in our diets—is a yellow dye used to add appealing color to pizza, orange soda, poultry, and cheese, among other things. Beta-carotene also gives carrots, peaches, apricots, and melons their natural color. And green, leafy vegetables contain the substance al-though the color is masked by chlorophyll. As chlorophyll is withdrawn from deciduous plants in the decreasing sunshine of autumn, beta-carotene shows through and is responsible for the leaves' yellow color.

Also present in the tanning pills is canthaxanthin, deep red-orange in color, found naturally in mushrooms, algae, cray-fish, and in the feathers of flamingos.

The tanning pills (with trade names such as Hawaiian Tan-ning Tablets, Tanamin, Oral-Tan, and Orobronze) have amounts of beta-carotene and canthaxanthin far in excess of that consumed in our diets, twenty to thirty times as much. The body is so overdosed with color that some of it accumulates in the skin. But not just skin is colored. Blood, sweat, feces, and urine may also take on an orange or red color, as do areas of the skin that are not pigmented, such as the palms of the hands and soles of the feet.

The Food and Drug Administration has not approved the sale or production of tanning pills in the United States, and they are thus illegal. Nevertheless, some health-food stores carry them and travelers to Canada and France, where they are legal, bring them back. A month's supply goes for twenty to thirty dollars. Side effects of the pills include nausea, cramps, and diarrhea, symptoms suffered by one-third of the subjects in a study at the Harvard Medical School.

To this point, I have focused primarily on the problems caused by using the various alternatives to sunlight, whether for seeing or for tanning. But therapeutic lights have also been developed for the treatment of disease.

PHOTOMEDICINE

In Chapter One it was shown how some physicians, in the name of control, began to replace sunlight with artificial lights in the treatment of certain diseases. Photomedicine has continued to the present and is successfully used to treat an assortment of skin diseases and other ailments. Among physicians, however, photomedicine is not held in uniformly high regard, and some blame it for causing as many problems as it treats. Such treatment-induced maladies are referred to as iatrogenic diseases wherein the proposed cure creates even greater problems than does the disease it treats. For example, cold sores (herpes simplex) were treated with light for several years until, in 1975, the Bureau of Radiological Health published a summary of its effectiveness. The treatment, they found, was not successful and furthermore, experimental studies showed that skin cells so treated could cause cancer when injected into animals.

Recent refinements in phototherapeutic techniques, however, have led to a renaissance in photomedicine. Currently, phototherapy with and without chemicals is used to treat pso-

riasis, vitiligo, polymorphous light eruption, hyperbilirubinism, and cancer, among other things.

Photomedicine works in three basic ways: by killing abnormal cells (called phototoxicity); by inducing a protective response in the skin (such as the stimulation of melanin production); and by altering certain metabolic substances in the body.

Phototoxicity

Ultraviolet radiation can be used to kill unwanted cells in
the body. In the treatment of psoriasis, patients are exposed
to UVA radiation, often after taking a psoralen drug orally.
Exposures of once or twice a month can clear the symptoms
and maintain psoriasis-free skin, but the treatments may have
undesirable side effects. Psoriasis maintenance treatments are
effective because UVA affects the immune system, possibly
by killing leukocytes. While the abnormal functioning of the
immune system is probably involved in the development of
psoriasis, depressing this system can have unwanted concom-
itants. Studies have shown that psoriasis patients undergoing
UVA therapy are more likely to develop skin cancers. The
possibility exists that they are induced by the therapy.

Surprisingly, some skin problems that are sunlight induced
can be treated with phototherapy. Polymorphous light eruptions become less severe after treatment with controlled doses
of UVR. Again, the abnormal cells which cause the problem
are killed.

Acne is also treated with UV phototherapy, but there is no
convincing evidence of its effectiveness. A tan may be induced through phototherapy and this can produce a cosmetically pleasing mask over the acne, but UVR does not reduce
the number of lesions. The often-observed phenomenon of
acne improving during the summer may be due to increased
blood flow to the face caused by heat. This may hasten the
healing of lesions.

Induction of a Protective Response

Some skin problems respond to phototherapy due to the
building of a protective response in the skin. As noted earlier,

UVA, in combination with a psoralen drug, can improve vitiligo. The psoralen is a photosensitizer, and skin develops pigment more readily when exposed to either sunlight or UVA lights. The polymorphous light eruptions also benefit from controlled doses of either artificial or natural light because the skin develops a protective layer of melanin. With a tan, the patient can better tolerate sunlight.

Alteration of Metabolites

Hyperbilirubinism, also called neonatal jaundice, also responds favorably to phototherapy. In this case, light acts directly on the bilirubin, altering it in such a way that it can be secreted in the urine. The liver's ability to metabolize toxins is also improved. Similarly, general pruritis, a symptom of chronic kidney failure, can be controlled by doses of UVB. Chemicals in the blood and skin are changed.

Optical Targeting

Perhaps the most exciting and general use of light in the treatment of disease lies in the area of optical targeting. John Parrish has outlined the procedure as follows:[7]

Visible light can pass through the epidermis of the skin, through the dermis, and even into the fat layers of the body. By choosing a particular wavelength, a therapist can determine precisely which layer of skin or underlying region will be affected by the light. Furthermore, drugs can be delivered to an exact site in the body by enveloping the medicine in a coating that breaks down when exposed to a certain wavelength of light. If, for example, we want a drug to exert its effect on a limited spot in the skin, say the dermis, the patient would be given the drug orally and after it is circulating in the body, the limited site would be irradiated by a beam of light of the particular wavelength that will release the drug.

Optical tageting is still in the developmental stage, but if Parrish's proposal works, side effects would be greatly reduced, since the drug would be inactive in parts of the body not so irradiated. We would have a pinpoint delivery system.

Photomedicine, if done carefully, seems to have a bright

future. Kendrick Smith, of the Stanford University School of Medicine, believes that if physicians were given formal training in the field of photobiology, a practice uncommon today, the potential hazards of the therapy would be minimized.[8] At one time phototherapy was used indiscriminately for all nature of problems, and adverse side effects were common. With more education of practitioners, phototherapeutic techniques may come to be a useful and safe way of effecting cures that we once appealed to the sun for.

Like sunlight, artificial light can be helpful or harmful. While it enables us to see and can be used therapeutically, there is ample reason to be cautious around it. Since our exposure to it has been recent, historically speaking, we know very little about its long-term consequences for human health. And as we have seen, artificial illumination is no substitute for the genuine article—sunlight.

APPENDIX:
THE SCIENTIFIC STUDY
OF LIGHT AND HEALTH

CONTEMPORARY research concerning the effects of sunlight and artificial light on human health bears little resemblance to similar endeavors conducted only a few decades ago. Unless familiar with the laboratory research of experimental biologists, you will probably be surprised to learn that most of our knowledge concerning light and health comes from research with nonhuman subjects. Rats, for example, are commonly used in laboratory studies. But by far the most popular subject for light research is a bacterium of the species Escherichia coli. How, you may ask, can bacteria so small that billions of them can be grown in a one-milliliter culture, tell us much about the effects of light on a 150-pound human?

Similarly, you may have been vexed to discover while reading this book, one which is ostensibly about *sunlight* and health, that much of the recent research was conducted indoors, usually in windowless labs. This sounds a little like measuring rainfall by putting a rain gauge under a leaky faucet. Can such unnatural setups ever tell us anything about the natural world outside?

Scientists have good reasons for working in laboratories. Only when they are able to control the conditions under which they work do they begin to obtain results that other scientists can replicate. Replication is the key to the acquisition of all scientific knowledge, for if one researcher discovers something that no one else can find, his or her results are suspect. Perhaps the study was done poorly and data collection was sloppy. More commonly, the conditions under which the observations were made may not have been duplicated by subsequent researchers.

Consider a research team that wants to find out how long it takes for a person to receive one MED while sitting in the sun. For another team to replicate this finding, thereby giving us increased confidence in the accuracy of the result, the first team had better provide complete information about the skin type of the subject, whether he had been in the sun prior to the experiment, what part of the body was irradiated, etc. Information such as this is usually provided in research reports. But the team had also better note that the study was conducted in Bangor, Maine, on June 21, that the sky was clear, that humidity was 55 percent of saturation, that ozone concentration was 80 percent of normal, that there was a moderate haze, that the closest tree was twenty feet away, ad infinitum. To be complete, they should also reveal the exact particulate matter that constituted the haze, and just how big that tree was.

Even after listing, were it possible, all of the relevant environmental data known to affect sunlight intensity, the nagging suspicion would remain—what if there are some unknown factors that moderate or augment the intensity of solar radiation? What if, indeed. Were the second research team to travel to Bangor on June 21 of the next year, verify that environmental conditions were (miraculously) similar, and conduct a replication study, results might still be quite dissimilar. Which team would you believe? There is no way of knowing.

If, on the other hand, the researchers were to irradiate a subject in the controlled environment of the laboratory with a given intensity of artificial light at a wavelength of 305 nm, researchers around the world could replicate this finding. They would smile at the reliability, or consistency, of the results.

But what results? Measuring something consistently doesn't mean that it is worth measuring or that it has relevance to anything in the real world. We will never find ourselves under a light emitting radiation only at 305 nm. What can this tell us about taking a walk in the sun? The question just posed involves the *external validity* of the research. We know that the laboratory findings can be repeated consistently, but should we care?

Perhaps. By conducting similar controlled studies with radiation at 297, 298, 299, 300 nm, all the way through the UV range, an action spectrum for erythema can be determined. A curve can be drawn which reveals the relationship between wavelength and MED, one that will show that it takes the least amount of time to get a sunburn at 297 nm and greater times at wavelengths larger and smaller. We now have something we can take outdoors. We can explain why we rarely burn at five o'clock on a sunny day though the sun is still hot—UVB is not present in the solar spectrum in any appreciable amount at this hour. And we now understand why people sunbathing at the Dead Sea report sunburns less frequently than heliophiles in other parts of the world—at this elevation radiation at and near 297 nm is negligible except for a short period of time around noon.

Applied research, the kind which tells us something that will benefit us in the real world, often progresses as follows: Someone observes an interesting relationship in the natural environment, say, that acne usually improves in the summertime. Laboratory research may show that a certain virus related to acne is killed by UVB radiation (a link, by the way, that is yet to be established). Dermatologists may consequently advise their acne patients to get outdoors for short periods a few days a week as part of their treatment, and the physicians may collect data on the success or failure of their advice. The course of study has progressed from the initial outdoor observation to a laboratory research program, and finally back outdoors again.

But extrapolating from the laboratory to the real world always presents troubling problems. We may learn that radiation at 297 nm most efficiently causes sunburns. Should we recommend that people avoid UVB? A dermatologist might say yes. An endocrinologist, aware of the importance of UVR which enters the eyes and benefits the whole organism, may say no.

A further problem of the laboratory is that, to the extent that it allows control of the phenomenon under study, it removes that phenomenon further and further from the natural phenomenon, which was the original inspiration for the research. Irradiating skin with successive wavelengths of monochromatic light, light of a single wavelength, can result in the establishment of an action spectrum for erythema. And since, by using animal subjects, we know that the action spectrum for skin cancer (squamous cell) is similar to the action spectrum for erythema, we may recommend that sensitive people avoid sunlight rich in UVB. What is not known is whether sunlight has wavelengths that *inhibit* the development of skin cancer. Visible light, as has recently been discovered, activates a process that repairs damage done by UVB. In short, there may be an interaction effect when different wavelengths strike the skin at the same time, with certain wavelengths causing damage and others promoting repair. To address this issue, laboratory studies would have to present different combinations of wavelengths throughout the spectrum. This is not feasible, since looking at all possible combinations of different wavelengths would require a number of studies that approaches what we commonly call infinity. Ultimately, the only satisfactory way to note the interactive effects of the various wavelengths present in sunlight is to use sunlight itself. We must often choose, then, between realism and control.

EXPERIMENTAL AND OBSERVATIONAL RESEARCH

In everyday parlance, any and all carefully conducted research is referred to as an experiment. In scientific circles, on the other hand, a firm distinction is drawn between research that adheres to the experimental method and that which is observational.

The Experiment

In true experimental research, scientists hold all the cards. They are the ones who decide where the research will be done (usually in the controlled environment of the laboratory), who the subjects will be, and what stimulus the subjects will be exposed to. Typically, experimenters divide subjects into two groups, the experimental group and the control group. Subjects are assigned randomly to either of these groups, random assignment helping to insure that there are no systematic differences, on the average, between experimental and control groups. Next, subjects in the experimental group are exposed to a stimulus, perhaps light of a given intensity. The control group is left alone. Scientists then check to see if there are any differences between experimental and control groups. If there are, the differences can be attributed to the experimental stimulus, the light.

Observational Research

In contrast, scientists doing observational research take their subjects where they find them. A dermatologist, for example, may question his skin cancer patients about their sunbathing habits and discover that the majority are sun worshippers. This can be taken as evidence that there is some connection between sunbathing and cancer of the skin. But the nature of the connection can only be guessed at. Perhaps overexposure to sunlight *causes* skin cancer, but the dermatologist's observational study cannot confirm this. For all the physician knows, the *majority* of Americans are sun worshippers. If, then, it is noted that skin cancer patients also tend to be sun worshippers, what has been demonstrated? Only that their habits do not differ materially from those without skin cancer. Of course, those sun worshippers who do not develop skin cancer never come to the attention of the physician since, in the absence of skin problems, they have better things to spend their money on than dermatologists.

Still, because we have so much evidence that sunlight does cause some types of skin cancer, it is tempting to believe that the observational study proves this. By itself it does not. All we can say is that it doesn't disprove it.

Cause and Effect

Let us consider another observational study, one that demonstrates the foolishness of thinking that this research proves cause and effect. In Australia, an investigator showed that people who wore hats were *more* likely to develop skin cancer than those who wore none. Did he conclude that the hats *caused* skin cancer, that straw has a toxic substance that creates neoplasms when in contact with the skin? Of course not. He concluded that people who had very fair complexions and suffered painful sunburns turned to hats for protection. Obviously, these are the very same people at high risk for skin cancer, and the increased incidence of it among them was to be expected. Skin type rather than hats is a more plausible explanation for the observed relationship. The hats were incidental. But note that this explanation, while appealing, has in no way been confirmed by the study itself. To show that skin type is a causative factor in the formation of skin cancer, an experiment would have to be conducted.

Individuals from each of the six skin types would be divided into experimental and control groups. The experimental groups would be instructed to sit out in the midday sun for two or three hours a day while the control groups would be kept indoors. If the experimental (sun-exposed) groups developed skin cancers at a higher rate than the control (nonexposed) groups, we could conclude that sunlight is a causative factor in skin cancer. Moreover, if sun-exposed subjects with skin types I and II developed skin cancer at a higher rate than sun-exposed subjects with skin types V and VI, skin type would be firmly implicated in cancer formation.

Unethical, you say, giving people skin cancer in the name of scientific research? Unethical, indeed, not to mention impractical. In humans, skin neoplasms sometimes develop decades after the harmful exposures are received. Few scientists are this patient, which brings us to our next point.

HUMAN AND NONHUMAN SUBJECTS

Because of ethical as well as practical considerations, scientists often turn to nonhuman organisms for their experimental subjects. As noted previously, the bacterium Escherichia coli is probably the most popular subject among light researchers. Other bacteria and bacterial viruses (known as phages) are also commonly used. For those who study the relationship between sunlight and *skin* damage, rats are often used, albino ones in particular because of their sensitivity to UVR.

Except among the antivivisectionists, the use of rats for research purposes answers the ethical problems. (The antivivisectionists stand mute on the question of bacterial subjects.) And many practical problems are addressed as well. Rats can develop skin cancers and other varieties of damage in weeks or months following irradiation rather than the decades that go by before humans manifest wrinkles, skin cancers, etc. Furthermore, some light-induced changes occur only rarely. While skin cancer rates are growing, it still only affects one in four hundred people in the United States during a given year. This means that prohibitively large populations of human subjects would be necessary before any experimental patterns could be observed. The difficulty in assembling large numbers of rats is smaller, and available bacterial subjects number in the billions. Thus, low-rate phenomena can be observed in large numbers.

But again there are trade-offs. While we solve some ethical and practical problems by using nonhuman subjects, we come up with some new questions. Do the reactions of bacteria and rats tell us anything about how we will react to light? The question of generalization looms. Earlier, we addressed the issue of generalizing knowledge gained in the laboratory to the world outside. Similarly, we must ask whether we can extrapolate results achieved with lower animals and microorganisms to humans.

In fact, we share much with rats and bacteria. We all have the same building blocks of life, the nucleic acids, though with different levels of complexity. And the nucleic acid DNA, which carries the genetic information that instructs cells how to grow, is common to all organisms. If light reaching the DNA of bacteria destroys or alters it, it will do the same in humans. (Provided, that is, that the light reaches a particular cell. Of all solar radiation, UVB has the greatest effect on DNA, and as we have seen, UVB penetrates only to the epidermis of the skin. Interior areas are thus protected. Bacteria, on the other hand, are so thin that all DNA is reached and affected by UVB.)

While we share much with all organisms, there are obviously some significant differences. Research on sun-induced retinal damage in rats may provide us with good information on the sequence of events that culminate in eye damage, but we are well advised not to extrapolate everything learned with rats to humans. If we find, for example, that daylight of normal intensity causes retinal damage in rats (and it does), we should not rush to enrich the sunglasses manufacturers. Instead, we should reflect that rats are nocturnal creatures and that their eyes are attuned to the low-level illumination of their nighttime activities. Humans, being diurnal, have eyes that are much less sensitive and that tolerate normal daylight intensities.

VARIED APPROACHES TO SCIENTIFIC RESEARCH

As you can see, no mode of research is perfect. That is why experimenters and observational researchers continue their studies without suffering crippling reservations concerning the utility of their work. As noted previously, research can move into, as well as out of, the laboratory, and the relevance of experimental work can be tested on the outside while the rigor of observational work can be put to the test in the laboratory. Though all individual research methods have their problems, different methods usually do not share the same problems and thus can be used to complement one another. The lack of control inherent in observational research is compensated for by the rigor of the experiment. And the sterility of the laboratory is compensated for when experimental findings are tested in the light of day.

NOTES

1: THERE GOES THE SUN

1. Norman E. Rosenthal, David A. Sack, J. Christian Gillin, Alfred J. Lewy, Frederick K. Goodwin, Yolanda Davenport, Peter S. Mueller, David A. Newsome, and Thomas A. Wehr, "Seasonal Affective Disorder," *Archives of General Psychiatry* 41 (1984): 72–80.

2. Valerie Beral, Helen Shaw, Susan Evans, and Gerald Milton, "Malignant Melanoma and Exposure to Fluorescent Lighting at Work," *Lancet*, (August 7, 1982): 290–93.

2: A SUNLIGHT PRIMER

1. National Research Council, "Environmental Impact of Stratospheric Flight: Biological and Climatic Effects of Aircraft Emissions in the Stratosphere," Climatic Impact Committee (Washington, D.C.: National Academy of Sciences, 1975).

2. National Academy of Sciences, "Causes and Effects of Stratospheric Ozone Reduction: An Update," (Washington, D.C.: National Academy Press, 1982).

3. Derek J. Cripps, "Natural and Artificial Photoprotection," *Journal of Investigative Dermatology* 76 (1981): 154–57.

4. Barry S. Paul and John A. Parrish, "The Interaction of UVA and UVB in the Production of Threshold Erythema," *Journal of Investigative Dermatology* 78 (1982): 371–74.

3: THE SUN AS FRIEND

1. Roy J. Shepard and S. Itoh, eds., *International Symposium on Circumpolar Health, 3rd* (Toronto: University of Toronto Press, 1976).

2. Richard J. Wurtman, "The Effects of Light on the Human Body," *Scientific American* 233 (1975): 68–77.

3. Ibid.

4. Ibid.

5. H. M. Hodkinson, P. Round, B. R. Stanton, and C. Morgan, "Sun-

light, Vitamin D, and Osteomalacia in the Elderly," *Lancet* 1 (April 28, 1973): 910–12.

6. Fritz Hollwich, *The Influence of Ocular Light Perception on Metabolism in Man and in Animal* (New York: Springer-Verlag, 1979).

7. Alfred J. Lewy, Thomas A. Wehr, Frederick K. Goodwin, David A. Newsome, and S. P. Markey, "Light Suppresses Melatonin Secretion in Humans," *Science* 210 (1980): 1267–69.

8. Alfred J. Lewy and David A. Newsome, "Different Types of Melatonin Circadian Secretory Rhythms in Some Blind Subjects," *Journal of Clinical Endocrinology and Metabolism* 56 (1983): 1103–7.

9. L. Wetterberg, "Melatonin in Humans: Physiological and Clinical Studies," *Journal of Neural Transmission*, Supplementum 13 (1978): 289–310.

10 N. Okudaira, D. F. Kripke, and J. B. Webster, "Naturalistic Studies of Human Light Exposure," *American Journal of Physiology* 254 (1983): R613–15.

11. See Hollwich.

12. Ibid.

13. Ibid.

14. Willy W. Avrach, "Climatotherapy at the Dead Sea," In *Psoriasis: Proceedings of the Second International Symposium*, E. M. Farber and A. J. Cox, eds. (New York: Yorke Medical Books, 1977).

15. John Ott, *Health and Light* (New York: Pocket Books, 1976).

16. See Beral et al.

17. N. M. Dantsig, D. N. Lazarev, and M. V. Sokolov, "Ultraviolet Installations of Beneficial Action," *Applied Optics* 6 (1967): 1872–76.

18. R. P. Feller, S. W. Burney, and I. M. Sharon, "Some Effects of Light on the Golden Hamster" (Paper presented at the meeting of the International Association for Dental Research, New York, N.Y., March 1970).

19. See Wurtman.

20. See Hollwich.

21. Edward E. Foulkes, *The Arctic Hysterias of the North Alaskan Eskimo* (Washington, D.C.: American Anthropological Association, 1972).

22. See Rosenthal et al.

23. Ibid.

24. Irving Geller, "Ethanol Preference in the Rat as a Function of Photoperiod," *Science* 173 (1971): 456–58.

4: THE SUN AS FOE

1. Wayne J. Anderson and Radames K. H. Gebel, "Ultraviolet Windows in Commercial Sunglasses," *Applied Optics* 16 (1977): 515–17.

2. Fred Hollows and David Moran, "Cataract—The Ultraviolet Risk Factor," *Lancet* 2 (December 5, 1981): 1249–50.

3. Reported in *New York Times*, September 3, 1982.

4. M. L. Kripke and M. S. Fisher, "Immunologic Parameters of Ultraviolet Carcinogenesis," *Journal of the National Cancer Institute* 57 (1976): 211–15.

5. B. H. R. Hill, "Immunosuppressive Drug Therapy as Potentiator of Skin Tumors in Five Patients with Lymphoma," *Australian Journal of Dermatology* 17 (1976): 16–18.

6. M. S. Kripke and M. L. Kripke, "Suppressor T Lymphocytes Control the Development of Primary Skin Cancers in Ultraviolet Irradiated Mice," *Science* 216 (1982): 1133–34.

7. D. Gordon and H. Silverstone, "Worldwide Epidemiology of Premalignant and Malignant Cutaneous Lesions, in *Cancer of the Skin,* R. Andrade, ed. (Philadelphia: W. B. Saunders, 1976), 405–55.

8. J. Scotto, T. R. Fears, and G. B. Gori, *Measurement of Ultraviolet in the United States and Comparisons with Skin Cancer Data* (Washington, D.C.: U.S. Department of Health, Education, and Welfare, Public Health Service, National Institutes of Health, 1975).

9. Walter Harm, *Biological Effects of Ultraviolet Radiation* (Cambridge: Cambridge University Press, 1980).

10. Robert G. Freeman, H. T. Hudson, and Robin Carnes, "Ultraviolet Wavelength Factors in Solar Radiation and Skin Cancer," *International Journal of Dermatology* 9 (1970): 232–35.

11. Knut Schmidt-Nielsen, *Desert Animals: Physiological Problems of Heat and Water* (New York: Dover Publications, Inc., 1979). For original study, see Edward F. Adolph, *Physiology of Man in the Desert* (New York: Hafner Publishing Co., 1947).

12. T. Stephen Jones, Arthur P. Liang, Edwin M. Kilbourne, Marie R. Griffin, Peter A. Patriarca, Steven G. Fite Wassilak, Robert J. Mullan, Robert F. Herrick, H. Denny Donnell, Keewhan Choi, and Stephen B. Thacker, "Morbidity and Mortality Associated with the July 1980 Heat Wave in St. Louis and Kansas City, Mo." *Journal of the American Medical Association* 247 (1982): 3327–31.

5: SENSE IN THE SUN

1. S. W. Tromp, *Medical Biometeorology* (New York: Elsevier Publishing Company, 1963).

2. Daniel S. Berger, "The Sunburning Ultraviolet Meter: Design and Performance," *Photochemistry and Photobiology* 24 (1976): 587–93.

3. Robert M. Sayre, Deborah L. Desrochers, Carol J. Wilson, and Edward Marlowe, "Skin Type, Minimal Erythema Dose (MED), and Sunlight Acclimatization," *Journal of the American Academy of Dermatologists* 5 (1981): 439–43.

4. A. V. J. Challoner, D. Corless, A. Davis, G. H. W. Deane, B. L. Diffey, S. P. Gupta, and I. A. Magnus, "Personnel Monitoring of Exposure

to Ultraviolet Radiation," *Clinical and Experimental Dermatology* 1 (1976): 175–79.

5. D. W. Owens, J. M. Knox, H. T. Hudson, and D. Troll, "Influence of Wind on Ultraviolet Injury," *Archives of Dermatology* 109 (1974): 200–201. Also see D. W. Owens, J. M. Knox, H. T. Hudson, and D. Troll, "Influence of Humidity on Ultraviolet Injury," *Journal of Investigative Dermatology* 64 (1975): 250–52.

6. Daniel S. Berger and Frederick Urbach, "A Climatology of Sunburning Ultraviolet Radiation," *Photochemistry and Photobiology* 35 (1982): 187–92.

7. A. Zweig and W. A. Henderson, Jr., "A Photochemical Mid-ultraviolet Dosimeter for Practical Use as a Sunburn Dosimeter," *Photochemistry and Photobiology* 24 (1976): 543–59.

8. C. D. J. Holman, I. M. Gibson, M. Stephenson, and B. K. Armstrong, "Ultraviolet Irradiation of Human Body Sites in Relation to Occupation and Outdoor Activity: Field Studies Using Personal UVR Dosimeters," *Clinical and Experimental Dermatology* 8 (1983): 269–77.

9. Frederick Urbach, John H. Epstein, and Donald P. Forbes, "Ultraviolet Carcinogenesis: Experimental, Global, and Genetic Aspects," in *Sunlight and Man*, ed. Thomas B. Fitzpatrick (Tokyo: University of Tokyo Press, 1974), 259–83.

10. Catherine Welsh and Brian Diffey, "The Protection Against Solar Actinic Radiation Afforded by Common Clothing Fabrics," *Clinical and Experimental Dermatology* 6 (1981): 577–82.

11. Kays H. Kaidbey and Albert M. Kligman, "An Appraisal of the Efficacy and Substantivity of the New High-potency Sunscreens," *Journal of the American Academy of Dermatologists* 4 (1981): 566–70.

12. John A. Parrish, Shukrallah Zaynoun, and R. Rox Anderson, "Cumulative Effects of Repeated Subthreshold Doses of Ultraviolet Radiation," *Journal of Investigative Dermatology* 76 (1981): 356–58.

13. S. W. Becker, Jr., and Victor E. Vaile, "Prevention of Solar Carcinoma," *Australian Journal of Dermatology*, 22 (1981): 56–58.

6: THE CIVILIZED SUN

1. See Wurtman.

2. See Hollwich.

3. John Hartung, "Light, Puberty, and Aggression," *Human Ecology* 6 (1978): 273–97.

4. See Beral et al.

5. John H. Epstein, "Suntan Salons and the American Skin," *Southern Medical Journal* 74 (1981): 837–40.

6. Peter Hersey, Enisa Hasic, Anne Edwards, Margot Bradley, Gregory Haran, and W. H. McCarthy, "Immunological Effects of Solarium Exposure," *Lancet* 1 (March 12, 1983): 545–48.

7. John A. Parrish, "New Concepts in Photomedicine: Photochemistry, Optical Targeting and the Therapeutic Window," *Journal of Investigative Dermatology* 77 (1981): 45–50.

8. Kendrick C. Smith, "Photobiology and Photomedicine: The Future Is Bright," *Journal of Investigative Dermatology* 77 (1981): 2–7.

FURTHER READING

Adolph, Edward F. *Physiology of Man in the Desert*. New York: Hafner Publishing Co., 1969 (originally published in 1947).

Alsop, G. F. "Health from Sunlight." *Literary Digest* 90 (August 28, 1926): 20.

Anderson, Wayne J., and Gebel, Radames K. H. "Ultraviolet Windows in Commercial Sunglasses." *Applied Optics* 16 (1977): 515–17.

"Are You a Heliophobe?" *Literary Digest* 106 (August 2, 1930): 30.

Avrach, Willy W. "Climatotherapy at the Dead Sea." In E. M. Farber and A. J. Cox, eds. *Psoriasis: Proceedings of the Second International Symposium*. New York: Yorke Medical Books, 1977.

Babbitt, Edwin D. *The Principles of Light and Color: The Classic Study of the Healing Power of Color*. Faber Birren, ed. Hyde Park, New York: University Books, 1967.

Batzell, Paul E. "Letting the Sun Cure Tuberculosis in Children." *Survey* 33 (1914): 102–4.

Becker, S. W., Jr., and Vaile, Victor E. "Prevention of Solar Carcinoma." *Australian Journal of Dermatology* 22 (1981): 56–58.

Beral, V., Shaw, H., Evans, S., and Milton, G. "Malignant Melanoma and Exposure to Fluorescent Lighting at Work." *Lancet*, (August 7, 1982): 290–93.

Berger, Daniel S. "The Sunburning Ultraviolet Meter: Design and Performance." *Photochemistry and Photobiology* 24 (1976): 587–93.

Berger, Daniel S., and Urbach, Frederick. "A Climatology of Sunburning Ultraviolet Radiation." *Photochemistry and Photobiology* 35 (1982): 187–92.

Bernhard, O. *Light Treatment in Surgery*. London, 1926.

Bickers, David R. "Position Paper—PUVA Therapy." *Journal of the American Academy of Dermatology* 8 (1983): 265–70.

Black, Homer S., Lenger, Wanda, Phelps, A. Warner, and Thornby, John I. "Influence of Dietary Lipid upon Ultraviolet-Light Carcinogenesis." *Nutrition and Cancer* 5 (1983): 59–68.

Blum, H. "The Physiological Effects of Sunlight on Man." *Physiological Reviews* 25 (1945): 453–83.

Boyce, P. R. *Human Factors in Lighting.* London: Applied Science Publishers, 1981.

Brady, William. "The Sun Cure." *Independent* 79 (1914): 414.

"A Careful Look into Tanning Booths." *The FDA Consumer* 14 (1980): 20–23.

Carmichael, L., and Dearborn, W. F. *Reading and Visual Fatigue.* Greenwood, Conn.: Greenwood Press, 1947.

Causes and Effects of Stratospheric Ozone Reduction: An Update. Washington, D.C.: National Academy Press, 1982.

Clayton, E. B. *Actinotherapy and Diathermy for the Student.* London: Bailliere, Tindall and Cox, 1945.

Coblenz, W. W., and Stair, R. "Evaluation of Ultraviolet Solar Radiation of Short Wave-lengths." *Journal of Research of the National Bureau of Standards* 16 (1936): 315.

Cripps, Derek J. "Natural and Artificial Photoprotection." *Journal of Investigative Dermatology* 76 (1981): 154–57.

Dantsig, N. M., Lazarev, D. N., and Sokolov, M. V. "Ultraviolet Installations of Beneficial Action." *Applied Optics* 6 (1967): 1872–76.

DeGruijl, F. R., Van Der Meer, J. B., and Van Der Leun, J. C. "Dose-Time Dependency of Tumor Formation by Chronic UV Exposure." *Photochemistry and Photobiology* 37 (1983): 53–62.

Denison, Charles. *Rocky Mountain Health Resorts.* Boston: Houghton, Osgood and Co., 1880.

"Differentiating Therapeutic Rays in Sunlight Cures." *Scientific American Monthly* 4 (1921): 199.

"Doctor Sun." *Literary Digest* 58 (August 31, 1918): 26.

Downes, A., and Blunt, T. P. "Researches on the Effect of Light upon Bacteria and Other Organisms." *Proceedings of the Royal Society of London* 26 (1877): 488–500.

East, Bion R. "Mean Annual Hours of Sunshine and the Incidence of Dental Caries." *American Journal of Public Health and the Nation's Health* 29 (1939): 777–80.

Ellinger, F. *Medical Radiation Biology.* Springfield, Ill.: Charles C Thomas, 1957.

Emerson, Haven. "Sunlight and Health." *American Journal of Public Health* 23 (1933): 437–40.

Epstein, John H. "Suntan Salons and the American Skin." *Southern Medical Journal* 74 (1981): 837–40.

Ewing, James. "Sunlight Cancer." *Ladies' Home Journal*, July 1941, pp. 28 and 36.

"Exposure to Sun Helps Prevent Cancer Deaths." *Science News Letter* 37 (1940): 198–99.

Fanselow, Dan L., Pathak, Madhu A., Crone, Margie A., Ersfeld, Dean A., Raber, Paul B., Trancik, Ronald J., and Dahl, Mark V. "Reusable

Ultraviolet Monitors: Design, Characteristics, and Efficacy." *Journal of the American Academy of Dermatology* 9 (1983): 714–23.

Fenner, Louise. "The Tanning Pill, a Questionable Inside Dye Job." *FDA Consumer* 16 (1982): 23–25.

Fisher, M. S., and Kripke, M. L. "Suppressor T Lymphocytes Control the Development of Primary Skin Cancers in Ultraviolet-Irradiated Mice." *Science* 216 (1982): 1133–34.

Fitzpatrick, T. B. *Sunlight and Man.* Tokyo: University of Tokyo Press, 1974.

Foulkes, Edward E. *The Arctic Hysterias of the North Alaskan Eskimo.* Washington, D.C.: American Anthropological Association, 1972.

Freeman, Robert G. "Data on the Action Spectrum for Ultraviolet Carcinogenesis." *Journal of the National Cancer Institute* 55 (1975): 1119–21.

Friedman, P. S. "Ultraviolet Carcinogenesis in Mice and Men." *British Journal of Dermatology* 109 (1983): 683–86.

Geller, Irving. "Ethanol Preference in the Rat as a Function of Photoperiod." *Science* 173 (1971): 456–58.

Giese, Arthur C. *Living with Our Sun's Ultraviolet Rays.* New York: Plenum Press, 1976.

Gilchrest, Barbara A., Szabo, George, Flynn, Evelyn, and Goldwyn, Robert M. "Chronologic and Actinically Induced Aging in Human Facial Skin." *Journal of Investigative Dermatology* 80 (1983): 81s–85s.

Gordon, D., and Silverstone, H. "Worldwide Epidemiology of Premalignant and Malignant Cutaneous Lesions." In *Cancer of the Skin,* edited by R. Andrade. Philadelphia: W. B. Saunders, 1976.

Harding, J. J. "Cataract: Sanitation or Sunglasses?" *Lancet* 1 (January 2, 1982): 39.

Harm, Walter. *Biological Effects of Ultraviolet Radiation.* Cambridge: Cambridge University Press, 1980.

Hartung, John. "Light, Puberty, and Aggression." *Human Ecology* 6 (1978): 273–97.

Hatfield, Elizabeth M. "Eye Injuries and the Solar Eclipse." *The Sight-saving Review* 40 (1970): 79–86.

Hedblom, E. E. "Snowscape Eye Protection." *Archives of Environmental Health* 2 (1961): 685.

"Heliotherapy: Miracles Wrought by Sunshine." *Review of Reviews* 50 (1914): 365–67.

Henderson, S. T. *Daylight and Its Spectrum.* 2nd ed. Bristol: Adam Hilger Ltd., 1977.

Hersey, P., Haran, G., Hasic, E., and Edwards, A. "Alteration of T Cell Subsets and Induction of Suppressor T Cell Activity in Normal Subjects After Exposure to Sunlight." *Journal of Immunology* 31 (1983): 171–74.

Hersey, P., Hasic, E., Edwards, A., Bradley, M., Haran, G., and Mc-

Carthy, W. H. "Immunological Effects of Solarium Exposure." *Lancet* 1 (March 12, 1983): 545–48.

High, A. S., and High, J. P. "Treatment of Infected Skin Wounds Using Ultra-violet Radiation—An *In Vitro* Study." *Physiotherapy* 69 (1983): 359–60.

Hill, B. H. R. "Immunosuppressive Drug Therapy as Potentiator of Skin Tumors in Five Patients with Lymphoma." *Australian Journal of Dermatology* 17 (1976): 16–18.

Hinsdale, Guy. "The Sun, Health, and Heliotherapy." *Scientific American Monthly* 9 (1919): 253–62.

Hodkinson, H. M., Round, P., Stanton, B. R., and Morgan, C. "Sunlight, Vitamin D, and Osteomalacia in the Elderly." *Lancet* 1 (April 28, 1973): 910–12.

Hollows, Fred, and Moran, David. "Cataract—The Ultraviolet Risk Factor. *Lancet* 2 (December 5, 1981): 1249–50.

Hollwich, Fritz. *The Influence of Ocular Light Perception on Metabolism in Man and in Animal.* New York: Springer-Verlag, 1979.

Holman, C. D. J., Gibson, I. M., Stephenson, M., and Armstrong, B. K. "Ultraviolet Irradiation of Human Body Sites in Relation to Occupation and Outdoor Activity: Field Studies Using Personal UVR Dosimeters." *Clinical and Experimental Dermatology* 8 (1983): 269–77.

Horwitz, David L. "Shedding Light on the Skin." *Comprehensive Therapy* 9 (1983): 3–4.

Huxley, Aldous. *The Art of Seeing.* New York: Harper & Brothers Publishers, 1942.

Jones, Billy M. *Health-seekers in the Southwest, 1817–1900.* Norman, Oklahoma: University of Oklahoma Press, 1967.

Kaidbey, Kays H., and Kligman, Albert M. "An Appraisal of the Efficacy and Substantivity of the New High-Potency Sunscreens." *Journal of the American Academy of Dermatologists* 4 (1982): 566–70.

Kantor, Jerry S. "Light as a Treatment for Nonseasonal Depression?" *American Journal of Psychiatry* 140 (1983): 1262.

Kime, J. W. "Sunlight a Necessity for the Maintenance of Health." *Scientific American Supplement* 81 (1916): 295.

Kligman, Lorraine H. "Intensification of Ultraviolet-Induced Dermal Damage by Infrared Radiation." *Archives of Dermatological Research* 272 (1982): 229–38.

Kripke, M. L., and Fisher, M. S. "Immunologic Parameters of Ultraviolet Carcinogenesis." *Journal of the National Cancer Institute* 57 (1976): 211–15.

Laurens, Henry. *The Physiological Effects of Radiant Energy.* New York: The Chemical Catalog Co., 1933.

Laurens, Henry. "Sunlight and Health." *Scientific Monthly* 42 (1936): 312–24.

Lee, John A. H. "Melanoma and Exposure to Sunlight." *Epidemiologic Reviews* 4 (1984): 110–36.

Lewy, Alfred J., Wehr, Thomas A., Goodwin, Frederick K., Newsome, David A., and Markey, S. P. "Light Suppresses Melatonin Secretion in Humans." *Science* 210 (1980): 1267–69.

Lewy, Alfred J., and Newsome, David A. "Different Types of Melatonin Circadian Secretory Rhythms in Some Blind Subjects." *Journal of Clinical Endocrinology and Metabolism* 56 (1983): 1103–7.

Lorand, Arnold. *Old Age Deferred.* Philadelphia: F. A. Davis Co., 1916.

Luce, Gay Gaer. *Biological Rhythms in Human and Animal Physiology.* New York: Dover Publications, 1971.

Luce, Gay Gaer. *Body Time: Physiological Rhythms and Social Stress.* New York: Pantheon Books, 1971.

Luckiesh, M., and Pacini, A. J. *Light and Health.* Baltimore: Williams & Wilkins Co., 1926.

McElroy, W. M., and Glass, B. *Light and Life.* Baltimore: Johns Hopkins Press, 1961.

MacKie, R. M., and Fitzsimons, C. P. "Risk of Carcinogenicity in Patients with Psoriasis Treated with Methotrexate or PUVA Singly or in Combination." *Journal of the American Academy of Dermatology* 9 (1983): 467–69.

Malinowski, Bronislaw. *Magic, Science, and Religion.* Garden City, New York: Doubleday & Company, 1948.

Maugh, Thomas H. II. "New Link Between Ozone and Cancer." *Science* 216 (1982): 396–97.

Maxwell, Kenneth J., and Elwood, J. Mark. "UV Radiation from Fluorescent Lights." *Lancet* 2 (September 3, 1983): 579.

Mayron, L., Mayron, E., Ott, J., and Nations, R. "Light, Radiation and Academic Achievement: Second Year Data." *Academic Therapy* 11 (1976): 397–407.

Mo, T., and Green, A. E. S. "A Climatology of Solar Erythema Dose." *Photochemistry and Photobiology* 20 (1974): 483–96.

"More Solar Surgery." *Literary Digest* 48 (June 13, 1914): 1432–33.

Morison, Warwick L. "Photoimmunology." *Journal of Investigative Dermatology* 77 (1981): 71–76.

National Research Council. *Environmental Impact of Stratospheric Flight: Biological and Climatic Effects of Aircraft Emissions in the Stratosphere.* Washington, D.C.: National Academy of Sciences, 1975.

Noell, W. K., and Albrecht, R. "Irreversible Effects of Visible Light on the Retina: Role of Vitamin A." *Science* 172 (1971): 76–80.

Okudaira, N., Kripke, D. F., and Webster, J. B. "Naturalistic Studies of Human Light Exposure." *American Journal of Physiology* 254 (1983): R613–15.

Ott, John. *Health and Light.* New York: Pocket Books, 1976.

Owens, D. W., Knox, J. M., Hudson, H. T., and Troll, D. "Influence of Wind on Ultraviolet Injury." *Archives of Dermatology* 109 (1974): 200–201.

Owens, D. W., Knox, J. M., Hudson, H. T., and Troll, D. "Influence of Humidity on Ultraviolet Injury." *Journal of Investigative Dermatology* 64 (1975): 250–52.

Palmer, Bruce. *Body Weather: How Natural and Man-made Climates Affect You and Your Health.* Harrisburg, Pa.: Stackpole Books, 1976.

Parrish, John A. "New Concepts in Photomedicine: Photochemistry, Optical Targeting and the Therapeutic Window." *Journal of Investigative Dermatology* 77 (1981): 45–50.

Parrish, John A. "Phototherapy and Photochemotherapy of Skin Diseases." *Journal of Investigative Dermatology* 77 (1981): 167–71.

Parrish, John A. "Ultraviolet Radiation Affects the Immune System." *Pediatrics* 17 (1983): 129–33.

Parrish, John A., Jaenicke, Kurt F., and Anderson, R. Rox. "Erythema and Melanogenesis Action Spectra of Normal Human Skin." *Photochemistry and Photobiology* 36 (1982): 187–91.

Parrish, John A., Zaynoun, Shukrallah, and Anderson, R. Rox. "Cumulative Effects of Repeated Subthreshold Doses of Ultraviolet Radiation." *Journal of Investigative Dermatology* 76 (1981): 356–58.

Pathak, Madhu A. "Sunscreens: Topical and Systemic Approaches for Protection of Human Skin Against Harmful Effects of Solar Radiation." *Journal of the American Academy of Dermatology* 7 (1982): 285–312.

Pathak, Madhu A., and Fanselow, Dan L. "Photobiology of Melanin Pigmentation: Dose/Response of Skin to Sunlight and Its Contents." *Journal of the American Academy of Dermatology* 9 (1983): 724–33.

Paul, Barry S., and Parrish, John A. "The Interaction of UVA and UVB in the Production of Threshold Erythema." *Journal of Investigative Dermatology* 78 (1982): 371–74.

Pearse, A. D., and Marks, Ronald. "Response of Human Skin to Ultraviolet Radiation: Dissociation of Erythema and Metabolic Changes Following Sunscreen Protection." *Journal of Investigative Dermatology* 80 (1983): 191–94.

Petersen, W. F. *Man, Weather, and Sun.* Springfield, Ill.: Charles C Thomas, 1947.

"A Physician's Protest Against the Current Delusion Of 'Glare.' " *Current Opinion* 62 (1917): 334.

Regan, J. D., and Parrish, J. A., eds. *The Science of Photomedicine.* New York: Plenum Press, 1980.

Relling, Mary V., and Dorr, Robert T. "Choosing a Sunscreen." *Arizona Medicine* 40 (1983): 550–54.

Robins, Perry, and Bennett, Richard G. *Current Concepts in the Management of Skin Cancer.* New York: Clinicom, Inc., 1978.

Rollier, Auguste. *Heliotherapy.* 2nd ed. London: Oxford University Press, 1927.

Rosenthal, Norman E., Sack, David A., Gillin, J. Christian, Lewy, Alfred J., Goodwin, Frederick K., Davenport, Yolanda, Mueller, Peter S., Newsome, David A., and Wehr, Thomas A. "Seasonal Affective Disorder." *Archives of General Psychiatry* 41 (1984): 72–80.

Roser-Maass, E., Hölzle, E., and Plewig, G. "Protection Against UV-B by UV-A-Induced Tan." *Archives of Dermatology* 118 (1982): 483–86.

Salterelli, C. G. "Light: The Forgotten Parameter." *Newsletter of the Center for Light Research* 6 (1977): 1.

Sams, W. Mitchell, Jr., Smith, J. Graham, and Burk, Peter G. "The Experimental Production of Elastosis with Ultraviolet Light." *Journal of Investigative Dermatology* 43 (1964): 467–71.

Sargent, Frederick II. *Hippocratic Heritage: A History of Ideas About Weather and Human Health.* New York: Pergamon Press, 1982.

Sayre, Robert M., Desrochers, Deborah L., Wilson, Carol J., and Marlowe, Edward. "Skin Type, Minimal Erythema Dose (MED), and Sunlight Acclimatization." *Journal of the American Academy of Dermatologists* 5 (1981): 439–43.

Schmidt-Nielsen, Knut. *Desert Animals: Physiological Problems of Heat and Water.* New York: Dover Publications, 1979.

Schreiber, Michael M., Moon, Thomas E., Meyskens, Frank L., and Mudron, Jean A. "Solar Ultraviolet Radiation and Skin Cancer: A Public Education Program." *Arizona Medicine* 40 (1983): 469–72.

Scotto, J., Fears, T. R., and Gori, G. B. *Measurement of Ultraviolet in the United States and Comparison with Skin Cancer Data.* Washington, D.C.: U.S. Department of Health, Education, and Welfare, Public Health Service, National Institutes of Health, 1975.

Sedrani, Saleh H., Elidrissy, Abdel Wahab T. H., and Arabi, Kamal M. El. "Sunlight and Vitamin D Status in Normal Saudi Subjects." *American Journal of Clinical Nutrition* 38 (1983): 129–32.

Setlow, Richard B. "DNA Repair, Aging, and Cancer." *National Cancer Institute Monograph* 60 (1982): 249–55.

Shepard, Roy J., and Itoh, S., eds. *Third International Symposium on Circumpolar Health.* Toronto: University of Toronto Press, 1976.

Shuster, Sam. "The Mechanism of Ultraviolet Erythema." *British Journal of Dermatology* 106 (1982): 235–36.

Slasson, E. E. "The Sun Cure." *Science Monthly* 16 (1923): 555–57.

Smith, D., Oei, T. P. S., Ng, K. T., and Armstrong, S. "Rat Self-Administration of Ethanol. Enhancement by Darkness and Exogenous Melatonin." *Physiology and Behavior* 25 (1980): 449–55.

Smith, John A., O'Hara, John, and Schiff, Anthony A. "Altered Melatonin Rhythms in Blind Men." *Lancet* 2 (October 24, 1981): 933.

Smith, Kendrick C. "Photobiology and Photomedicine: The Future Is Bright." *Journal of Investigative Dermatology* 77 (1981): 2–7.

Smith, Wesley D. "Suntans: Good News and Bad." *Sciquest* 52 (1979): 10–13.

"Substitute-Sunshine For Miners." *Literary Digest* 104 (January 25, 1930): 34 and 37.

"The Sun as Surgeon." *Literary Digest* 48 (February 14, 1914): 427.

Sutherland, Betsy M. "Photoreactivation." *Bioscience* 31 (1981): 439–44.

Swan, Herbert S., and Tuttle, George W. "Planning Sunlight Cities." *The American City* 17 (September 1917): 213–17.

Thorington, Luke. "Actinic Effects of Light and Biological Implications." *Photochemistry and Photobiology* 32 (1980): 117–29.

Tromp, S. W. *Medical Biometeorology.* New York: Elsevier Publishing Co., 1963.

Tromp, S. W., and Faust, V. "Influence of Weather and Climate on Mental Processes in General and Mental Diseases in Particular." In S. W. Tromp, ed. *Progress in Biometeorology.* Div. A, vol. 1, pt. II, period 1963–1975. Amsterdam: Swets & Zeitlinger, 1977.

Tromp, S. W., and Weihe, W. H., eds. *Biometeorology.* Vol. 4, pt. I. Amsterdam: Swets & Zeitlinger, 1970.

Urbach, Frederick. "Causes and Effects of Stratospheric Ozone Reduction: An Update." *Journal of the American Academy of Dermatology* 7 (1982): 217–73.

Urbach, Frederick, ed. *The Biologic Effects of Ultraviolet Radiation (with Emphasis on the Skin).* New York: Pergamon Press, 1969.

Wade, Nicholas. "Too Much Light Shed on Body Public." *Science* 206 (1979): 913.

Watson, Alan. "Sunscreen Effectiveness: Theoretical and Practical Considerations." *Australian Journal of Dermatology* 24 (1983): 17–22.

Welsh, Catherine, and Diffey, Brian. "The Protection Against Solar Actinic Radiation Afforded by Common Clothing Fabrics." *Clinical and Experimental Dermatology* 6 (1981): 577–82.

Wetterberg, L. "Melatonin in Humans: Physiological and Clinical Studies." *Journal of Neural Transmission,* Supplementum 13 (1978): 289–310.

Wever, R. A., Polasek, J., and Wildgruber, C. M. "Bright Light Affects Human Circadian Rhythms." *Pflügers Archiv* 396 (1983): 85–87.

Wiemer, D. Robert, and Spira, Melvin. "Ultraviolet Light and Hyperpigmentation in Healing Wounds." *Annals of Plastic Surgery* 11 (1983): 328–30.

Wurtman, Richard J. "The Effects of Light on the Human Body." *Scientific American* 233 (1975): 68–77.

Wyke, P. L. "Skin Cancer and the Ultraviolet Spectrum." *Lancet* 2 (December 10, 1983): 1372–73.

Young, Richard W. "Visual Cells, Daily Rhythms, and Vision Research." *Vision Research* 18 (1978): 573–78.

Zigman, Seymour. "The Role of Sunlight in Human Cataract Formation." *Survey of Ophthalmology* 27 (1983): 317–25.

INDEX

Acne, 194
ACTH, 183-84, 188
Alcohol, 81
Alcoholism, 98, 100-101
Animals
  and color change, 69
  and sexual behavior, 93-94
  and sexual development, 76, 93
Antibiotics, 34
Antigens, 114
Apollo, 8
Architecture, 25-26, 186-87.
    See also Sunlight cities.
Arctic Circle, 64, 88
Arctic hysteria, 97-98
Arthritis, 85
Artificial light
  and ACTH level, 183-84
  and cortisol level, 183-84
  and eyestrain, 183
  and melanoma, 184-86
  and visual processes, 183
    See also Fluorescent lights, In-
    candescent lights.
Asthenia, 70-71
Atmosphere
  primitive, 44
  and solar spectrum, 43-44
Aztec Indians, 12

Babbitt, Edwin, 30
Bacteria, 19
Basal-cell carcinoma, 115-16
  and sunbathing, 118
  treatment of, 121-22
    See also Skin cancer.
Beral, Valerie, 184-86
Berger, Daniel, 142
Beta-carotene, 192
Bird migration, 93
Black lung disease, 67
Blacks
  and skin type, 59
  and vitamin D deficiency, 135
Blindness, 68
  and blood cholesterol, 80
  and blood platelets, 80
  and the endocrine system, 67
  and melatonin rhythms, 77
  monetary compensation for, 82
  and testosterone level, 95
  and white blood cell count, 80
    See also Cataracts.
Blood
  cholesterol, 80
  hemoglobin, 79
  platelets, 80
  protein, 79-80
  red cells, 79

Blood, *(continued)*
  white cells, 80
Blood pressure, 20

Calcium
  and Arctic hysteria, 97-98
  and bones, 70
  circadian rhythm of, 97-98
  and emotional problems, 96-98
  and rickets, 21
    See also Vitamin D.
Cancer, 91
  sunlight and prevention of, 38
  treatment with sunlight, 86-87
    See also Skin cancer.
Cataracts, 32, 83, 105-106
    See also Eyes.
Challoner, A.V.J., 155
Cholesterol. *See* Blood, cholesterol.
Circadian rhythms, 97-98
Climatotherapy
  decline of, 19
  and tuberculosis, 17-19
  and westward migration, 18
Clothing, 165-66
Clouds
  and skylight, 141-42
  and UVR, 140, 142
Collagen, 110
Color therapy, 29-30
Cortisol, 77-78, 183-84, 188
Criminality, 100

Dead Sea, 84-86
Dehydration, 129-30
Dennison, Charles, 17
Depression
  "gray sky syndrome," 99
  historical cures of, 98
  and melatonin, 98
  Seasonal Affective Disorder (SAD),
    99-100
Dermatitis, 31-32, 66
DNA
  destruction of, 204
  repair of, 38
  and skin cancer, 113-14
Dominant oculocutaneous hypomel-
    anism (DOH), 123

Dosimeters, 155
  and field studies, 161-63

Elastin, 110
Elderly, the
  and calcium deficiency, 70
  and life extension, 25
  and heat stroke, 131
  and hyperbilirubinism, 83
  and nutritional osteomalacia, 72
  and Vitamin D deficiency, 72
Erythropoietic photoporphyria, 123
Escherichia coli, 197, 203
Eskimos, 65.
    See also Arctic Circle.
Eyes, 90-91
  and damage from infrared radiation,
    106
  and damage from UVR, 32, 103-6
  and sunscreens, 171
    See also Cataracts; Vision.

Finsen, Niels, 19-20
Fluorescent lights, 179
  and melanoma, 87, 184-86
Full-spectrum light, 90-92
  and ACTH level, 188
  and architecture, 186
  and cancer, 91
  and corrective lenses, 92, 188
  and cortisol production, 188
  and illness, 90
  and plant growth, 90

Global radiation, 141
Gonads, 181

Hats, 37, 166, 202
Heart, 82
Heat stroke, 130-31
Heliobalneotherapy, 85
Heliophobia, 33
Helios, 8
Heliotherapy, 35
  in ancient Greece, 15-16
  in ancient Rome, 16
  eighteenth century, 17
  Finsen's, 19-20
  nineteenth century, 29

Rollier's, 21-23
in the U.S., 24-25
    See also specific diseases, e.g.,
    Cancer, Psoriasis, etc.
Hemoglobin. *See* Blood, hemoglobin.
Herpes simplex, 193
Hippocrates, 15
Hollwich, Fritz, 74-75, 80, 82, 95, 181,
    184, 188
Hopi Indians, 12-13
Humidity
    and skylight, 141
    and sunburning, 158
Hyperbilirubinism, 83, 195

Immune system
    and skin cancer, 114-15
    and tanning salons, 191-92
    and UVR, 114
Incandescent lights, 179
Infrared radiation, 47-48
    and clothing, 165-66
    definition of, 47
    and eye damage, 106
    and flushing of skin, 55
    and heat stroke, 130-31
Insomnia, 183

Jet lag, 77

Keratosis, actinic, 110
Kripke, Margaret, 115

Lewy, Alfred, 39, 76, 99
Liver, 81, 83

Maya Indians, 11-12
Melanin, 173
Melanoma, 116-17
    and bathing suits, 119
    and fluorescent lights, 87, 184-86
    and sunlight, 118-19
    treatment of, 121-22
    See also Skin cancer.
Melatonin
    and alcohol consumption, 100-101
    circadian rhythm of, 77-78
    and cortisol production, 77-78
    and depression, 39, 98

and jet lag, 77
and light, 76-77
and sexual behavior, 39, 95
Miners, 65, 66
Myths, 7

National Academy of Sciences, 50-51
Neonatal jaundice. *See* Hyperbiliru-
    binism.
Nutritional osteomalacia, 72
Nystagmus, 66

Optical targeting, 195
Ott, John, 86-87, 90-91
Ozone, 44, 48-51, 138
    and Freons, 37, 50
    and nuclear explosions, 50-51
    reduction of, 50-51
    and SSTs, 37, 50
    and UVB, 48, 49

PABA, 168-70
PABA derivatives, 168-70
Parrish, John, 61-62, 176-77
Photomedicine
    and alteration of metabolites, 196
    and optical targeting, 195
    and phototoxicity, 194
    problems with, 31, 193
    and protective response, 194
    and psoriasis, 194
    See also Phototherapy.
Photosensitivity, 123-27
    and consumer products, 125
    and drugs, 124-25, 126-27
    and plants, 126
Phototherapy, 30-31
    of miners, 66
    See also Photomedicine.
Phototoxicity, 194
Pineal gland, 75-76, 94
Platelets. *See* Blood, platelets.
Pollution, 51, 159-60
    natural, 53
Polymorphous light eruptions, 194, 195
Porphyria, 126
Pott's disease, 24
Protein. *See* Blood, protein.

Psoralens
  natural, 126
  and tanning, 178
  and treatment of psoriasis, 84, 194
  and treatment of vitiligo, 83-84
Psoriasis, 84-85, 194
Puberty, 181
Pueblo Indians, 12

Rainbows, 45
Red blood cells. See Blood, red cells.
Reflection, 155-157
Research
  experimental, 200-202
  observational, 201-2
  replication of, 198
Rickets, 21, 71, 73.
  See also Vitamin D.
Rollier, Auguste, 21-23
Rosenthal, Norman, 99
Russia. See USSR.

Seasonal Affective Disorder (SAD).
  See Depression, Seasonal Affective
  Disorder (SAD).
Sexual behavior
  of animals, 93-94
  and artificial light, 181-82
  and circadian rhythms, 94
  and melatonin, 95
  of miners, 66
  seasonality of, 95-96
  and sunlight, 38
Sexual development
  of animals, 76, 93
  and artificial light, 181
  and menstruation, 94
Shade
  and protection from UVR, 155
  and skylight, 166-67
Skin, 54
  dry, 109
  and UVA, 54, 55, 56
  and UVB, 54, 55, 56
Skin cancer, 111-24
  and bathing suits, 33
  early laboratory studies of, 31

and lifetime exposure, 163-64
public awareness of, 35-36
and single dose of UVR, 40
treatment of, 121-22
  See also Basal-cell carcinoma,
  Melanoma, and Squamous-cell
  carcinoma.
Skin color, 57, 134
Skin type, 57-59, 111
Sky, 45-46
Skylight
  and clouds, 141-42
  and shade, 155, 166
Smog, 51-52.
  See also Pollution.
Snow
  and reflection of UVR, 105
Squamous-cell carcinoma, 116
  and sunlight, 118
  treatment of, 121-22
  See also Skin cancer.
Stonehenge, 9-10
Suberythemal dose, 174
Suicide, 96, 98
Sun
  description of, 42
  eclipse of, 9, 42, 106
  energy from, 42-43
  spectrum of, 43
Sunbathing
  in California, 24
  and safety, 174
  and UVR, 162-63
Sunburning, 20, 55-56, 107
  effects of repeated, 108-9
  and heat, 158
  and humidity, 158
  skin tests to determine, 160-61
  treatment of, 108
  and UVA, 61
  and UVB, 55
  and wind, 159
Sunburning Ultraviolet Meter, 142-43
Sunglasses, 87, 103-5
  problems with, 91
  and protection, 32
Sunlamps, 188-89
Sunlight cities, 26
"Sunlight starvation," 88

Sun Protection Factor (SPF), 168
    field studies of, 169
    and skin type, 59
Sunrise, 46
Sunblocks, 167, 168
Sunscreens
    early, 33, 167
    and eye irritation, 171
    problems with, 171-72
    and skin cancer, 168
    and skin type, 171
    and Sun Protection Factor (SPF), 168
    and sweating, 169-70
    and washoff, 170
        See also PABA; PABA deriva-
        tives.
Sunset, 46
Suntan. See Tanning.
Sun worship
    in the Americas, 11-13
    in ancient Egypt, 8
    in ancient Greece, 8
    in ancient Rome, 9
    and Christianity, 10-11
Sweating, 169, 171

Tanning, 56, 175-78
    with chemicals, 192-93
    immediate, 107
    and protection, 172-73
    and UVA, 61
Teeth, cavities of, 65, 71, 89
Temperature, 128-31
Temperature inversion, 52
Testosterone, 95, 181
T-lymphocytes, 114
Tuberculosis
    of the bone, 22-23, 24
    and climatotherapy, 17-19
    of the lungs, 17-19, 34-35, 65
    of the skin, 20

Ultraviolet-A (UVA), 60-61, 104
    definition of, 47
    and skin cancer, 120
    and sunburning, 61-62
    and tanning, 61
    and treatment of psoriasis, 84
    and wrinkling, 61

See also Ultraviolet Radiation
    (UVR).
Ultraviolet-B (UVB), 60-62
    definition of, 47
    and ozone, 50
    and skin cancer, 60
    and sunburning, 55
    and tanning, 56
        See also Ultraviolet Radiation
        (UVR).
Ultraviolet-C (UVC), 50, 60
    definition of, 47
        See also Ultraviolet Radiation
        (UVR).
Ultraviolet Intensity Tables, 244-54
Ultraviolet Radiation (UVR)
    and altitude, 139-40
    and latitude, 139-40
    scattering of, 46
    and season, 136-38
    and time of day, 138
        See also Ultraviolet-A (UVA); Ul-
        traviolet-B (UVB); and Ultravi-
        olet-C (UVC).
Urbach, Frederick, 163-64
Urocanic acid, 169
USSR, 66, 88-89

Vision, 53-54, 74, 183.
    See also Eyes.
Vitamin D, 21, 35, 70-73, 134, 171
    and blacks, 135
    and the elderly, 70, 72-73
    synthetic, 71
    and toxicity of, 127-28.
    See also Rickets.
Vitiligo, 83-84, 178

Water
    and reflection of UVR, 156-57
Wetterberg, L., 77
White blood cells. See Blood, white
    cells.
Wounds, 20
Wrinkling, 109-110, 173
    and UVA, 61
Wurtman, Richard, 93, 180

Xeroderma pigmentosum (XP), 123

ABOUT THE AUTHOR

Michael J. Lillyquist, a native of New York State, was for twelve years a professor of psychology at the University of Wisconsin at Platteville. His interest in sunlight and health began when he moved to Tucson to become a freelance writer and was diagnosed as having skin cancer. He knew he would have to learn a great deal more about the effects of the sun if he was to safely enjoy the benefits of sunny Arizona. *Sunlight and Health* was the result.